计算机科学与技术丛书

Java多线程与线程池技术详解

肖海鹏　牟东旭 ◎ 著
Xiao Haipeng　Mou Dongxu

JAVA MULTITHREADING AND THREAD POOL TECHNOLOGY

清华大学出版社
北京

内 容 简 介

本书全面、系统地讲解了多线程和线程池技术。

全书共分为 10 章，第 1~3 章为基本的多线程技术基础，包含创建线程、线程状态转换、线程间同步等内容。第 4~8 章为线程池技术，包含自定义线程池、通过 Executors 工具类创建线程池、线程池的底层阻塞队列结构、重入锁等内容。第 9 章和第 10 章为多线程技术的应用部分，包含基于 Tomcat 的线程池应用、基于 ThreadLocal 的数据库连接管理、生成唯一的订单号等内容。

全书提供了大量应用实例，每章后面均附有习题。

本书适合作为高等院校计算机、软件工程专业高年级本科生、研究生的教材，同时可供有一定编程经验的软件开发人员、广大科技工作者和研究人员在进行多线程编程时参考使用。

本书封面贴有清华大学出版社防伪标签。无标签者不得销售。
版权所有，侵权必究。举报：010-62782989，beiqinquan@tup.tsinghua.edu.cn。

图书在版编目（CIP）数据

Java 多线程与线程池技术详解 / 肖海鹏，牟东旭著. —北京：清华大学出版社，2021.4
（计算机科学与技术丛书）
ISBN 978-7-302-57373-9

Ⅰ. ①J⋯　Ⅱ. ①肖⋯ ②牟⋯　Ⅲ. ①JAVA 语言–程序设计　Ⅳ. ①TP312.8

中国版本图书馆 CIP 数据核字（2021）第 017873 号

责任编辑：刘　星
封面设计：吴　刚
责任校对：李建庄
责任印制：杨　艳

出版发行：清华大学出版社
网　　址：http://www.tup.com.cn, http://www.wqbook.com
地　　址：北京清华大学学研大厦 A 座　　邮　　编：100084
社　总　机：010-62770175　　邮　　购：010-83470235
投稿与读者服务：010-62776969，c-service@tup.tsinghua.edu.cn
质　量　反　馈：010-62772015，zhiliang@tup.tsinghua.edu.cn
课　件　下　载：http://www.tup.com.cn,010-83470236

印　装　者：三河市金元印装有限公司
经　　销：全国新华书店
开　　本：186mm×240mm　　印　张：20.5　　字　数：461 千字
版　　次：2021 年 5 月第 1 版　　印　次：2021 年 5 月第 1 次印刷
印　　数：1~2500
定　　价：79.00 元

产品编号：087016-01

前言
FOREWORD

一、为什么要写本书

1965 年，戈登·摩尔发现了一个惊人的集成电路发展趋势：当价格不变时，集成电路上可容纳的元器件的数目，每 18～24 个月会增加一倍，性能也将提升一倍。如果这个趋势继续，计算能力相对于时间周期将呈指数式的上升，这被称为摩尔定律。

50 多年过去了，摩尔定律目前仍然没有被打破，但是马上就要面临巨大的物理瓶颈。现在的最好的芯片技术是台积电的 5nm 工艺，下一个目标是 3nm，而硅原子在 1nm 空间只能排列三四个。这对于依赖硅为半导体材料的芯片产业而言，如果没有新的半导体材料出现，其很快就将面临物理极限挑战。

多核处理器技术是硬件发展的另一大趋势。通过使用 CPU 多核架构，可以有效缓解单个芯片运算能力不足带来的尴尬。现在几乎所有的笔记本计算机、台式计算机，还有服务器等普遍使用了多核处理器。多核 CPU 的普及要求系统软件和应用软件架构必须要与时俱进。因此，掌握并发编程技术，开发出适用于多核处理器的并发程序，充分利用 CPU 的并发处理优势，是对所有高级软件开发人员的硬性要求。

目前，Java 平台在服务器端的后台编程中占据着无法动摇的统治地位，如典型的 Web 服务器就是一个允许几十万人同时访问的高并发环境。因此 Java 的高并发编程是每一个 Java 编程人员必须要掌握的核心知识。

高并发处理不仅在 Web 服务器上，在应用服务器、数据库服务器、大数据服务器等服务器上，同样都是高并发环境，当然在不同服务器上的软件架构是完全不同的。

多线程的并发编程从来都不是一件容易的事情，即使对高级软件架构人员也是如此。Java 平台关于多线程部分的 API，历经多次频繁地调整，始终无法稳定下来，死锁和 CPU 利用率不高一直困扰着这些高级架构师。直到 2004 年，大神级的 Doug Lea 横空出世，在 JDK5 中推出了 java.util.concurrent 开发包，这才使 Java 彻底稳住了阵脚。

Java 多线程基础库稳定后，服务器并发编程技术发展迅速，目前基于 Java EE 8 规范的服务器性能更加优秀。本书中包含的所有案例均基于 JDK1.8 版本并已调试成功。

二、内容特色

与同类书籍相比，本书有如下特色。

1. 案例生动易懂，读者入门容易

国外关于多线程编程的经典图书，大多晦涩难懂，让人看后感觉云山雾罩、一头雾水；国内多线程相关书籍又过于浅显。因此，市场上没有一本让多数读者感觉满意的多线程图书，尤其是高校本科生和研究生，想选择一本多线程教材是非常困难的。本书中的大量案例，都贴合实际生活，生动鲜活，容易记忆，容易理解，读者可以轻松上手。

2. 原理透彻，注重应用

本书对多线程相关理论分门别类、层层递进地进行了详细的叙述和透彻的分析，既体现了各知识点之间的联系，又兼顾了其渐近性。本书在介绍每个知识点时都给出了该知识点的应用场景，同时配合源代码分析；本书真正体现了理论联系实际的理念，使读者能够体会到"学以致用"的乐趣。

3. 配套资源，超值服务

本书提供以下相关配套资源：

- 书中涉及的程序代码、习题答案，可以关注"人工智能科学与技术"微信公众号，在"知识"→"资源下载"→"配书资源"菜单获取下载链接（也可以到清华大学出版社网站本书页面下载）。
- 本书有完整的配套视频，可以到 CSDN、51CTO、腾讯课堂等网站观看，最简单的入口就是在百度直接搜索：肖海鹏，即可找到各网站的视频入口。在配套资源中也有视频观看链接（注：配套视频请读者自行购买观看）。

三、结构安排

本书第 1~3 章为多线程基础知识，第 4~8 章为线程池技术，第 9 章和第 10 章为多线程技术应用。

四、读者对象

- 电子信息工程、计算机科学与技术相关专业的本科生、研究生；
- 所有 Java 软件开发人员。

五、致谢

限于编者的水平和经验，加之时间比较仓促，疏漏之处在所难免，敬请读者批评指正，有兴趣的朋友可发送邮件至 workemail6@163.com。

编者
2021 年 3 月于北京

目录
CONTENTS

第 1 章　多线程基础 ··· 1

  1.1　进程与线程 ·· 1

    1.1.1　操作系统与进程 ·· 1

    1.1.2　线程与进程的关系 ·· 2

  1.2　多线程启动 ·· 3

    1.2.1　线程标识 ··· 5

    1.2.2　Thread 与 Runnable ·· 6

    1.2.3　run()与 start() ·· 7

    1.2.4　Thread 源码分析 ··· 9

  1.3　线程状态 ·· 11

    1.3.1　NEW 状态 ··· 11

    1.3.2　RUNNABLE 状态 ··· 12

    1.3.3　BLOCKED 状态 ··· 13

    1.3.4　WAITING 状态 ··· 14

    1.3.5　TIMED_WAITING 状态 ·· 19

    1.3.6　WAITING 与 BLOCKED 的区别 ·· 21

    1.3.7　TERMINATED 状态 ··· 21

    1.3.8　线程状态转换 ·· 22

  1.4　sleep()与 yield() ·· 22

    1.4.1　线程休眠 sleep() ·· 22

    1.4.2　线程让步 yield() ·· 24

  1.5　线程优先级 ·· 25

    1.5.1　线程优先级与资源竞争 ··· 26

    1.5.2　案例：大型浮点运算测试 ··· 26

    1.5.3　案例：多线程售票 ·· 28

  1.6　守护线程 ·· 30

    1.6.1　守护线程的概念 ·· 30

	1.6.2	案例：清道夫与工作者	31
1.7		本章习题	34

第 2 章 线程安全与共享资源竞争 37

2.1		synchronized 同步介绍	37
2.2		synchronized 同步方法	38
	2.2.1	同步方法调用流程	41
	2.2.2	同步方法之间的互斥	41
	2.2.3	同步方法与非同步方法	44
2.3		synchronized 同步静态方法	46
	2.3.1	单例高并发问题	46
	2.3.2	类锁与对象锁	47
	2.3.3	静态同步方法之间互斥	47
	2.3.4	静态同步方法与静态非同步方法	49
2.4		synchronized 同步代码块	50
	2.4.1	锁当前对象	50
	2.4.2	锁其他对象	51
	2.4.3	锁 Class	51
2.5		项目案例：火车售票	52
	2.5.1	共享任务模式	52
	2.5.2	多任务模式	55
	2.5.3	共享车票资源	56
2.6		项目案例：家庭消费	59
2.7		项目案例：别墅 Party	63
	2.7.1	无锁模式	63
	2.7.2	单锁模式	66
	2.7.3	双锁模式	67
2.8		JDK 常见类的线程安全性	69
	2.8.1	集合 ArrayList 与 Vector	69
	2.8.2	StringBuffer 与 StringBuilder	71
	2.8.3	HashMap 与 ConcurrentHashMap	72
2.9		本章习题	73

第 3 章 多线程通信 76

3.1		wait()与 notify()	76
	3.1.1	阻塞当前线程	76

	3.1.2 案例分析：厨师与侍者 1	78
	3.1.3 案例分析：厨师与侍者 2	81
	3.1.4 案例分析：两个线程交替输出信息	85
3.2	join 线程排队	87
	3.2.1 加入者与休眠者	87
	3.2.2 案例：紧急任务处理	89
	3.2.3 join 限时阻塞	91
3.3	线程中断	93
	3.3.1 中断运行态线程	93
	3.3.2 中断阻塞态线程	95
	3.3.3 如何停止线程	97
3.4	CountDownLatch 计数器	98
3.5	CyclicBarrier 屏障	100
	3.5.1 案例：矩阵分行处理	101
	3.5.2 案例：赛马游戏	103
3.6	Exchanger	104
3.7	Semaphore 信号灯	107
3.8	死锁	108
	3.8.1 案例：银行转账引发死锁	109
	3.8.2 案例：哲学家就餐死锁	112
3.9	本章习题	115

第 4 章 线程池入门 117

4.1	ThreadPoolExecutor	117
	4.1.1 创建线程池	118
	4.1.2 关闭线程池	121
4.2	Executor 接口	123
4.3	ExecutorService 接口	124
	4.3.1 Callable 返回任务执行结果	125
	4.3.2 shutdown 与 shutdownNow	127
4.4	Executors 工具箱	127
	4.4.1 newCachedThreadPool	128
	4.4.2 newFixedThreadPool	133
	4.4.3 newSingleThreadExecutor	137
	4.4.4 newScheduledThreadPool	141
	4.4.5 newWorkStealingPool	143

4.5 线程工厂与线程组···151
  4.5.1 线程组··151
  4.5.2 线程与线程组···152
  4.5.3 线程工厂接口···155
  4.5.4 默认线程工厂实现··156
  4.5.5 线程池与线程工厂··157
4.6 线程池异常处理···158
  4.6.1 异常捕获···158
  4.6.2 UncaughtExceptionHandler 处理异常··158
  4.6.3 Future 处理异常··161
4.7 本章习题··163

第 5 章 线程池与锁···165

5.1 重入锁 ReentrantLock···165
  5.1.1 重入锁··166
  5.1.2 互斥锁··167
  5.1.3 ReentrantLock 与 synchronized···169
  5.1.4 尝试加锁并限时等待···171
5.2 重入锁与 Condition···173
  5.2.1 案例分析：厨师与侍者··176
  5.2.2 案例分析：缓冲区队列··178
5.3 读锁与写锁···181
  5.3.1 案例：并发读写集合···182
  5.3.2 案例：Map 并发控制··193
  5.3.3 数据库事务与锁···197
5.4 公平锁与非公平锁··198
5.5 本章习题··201

第 6 章 线程池与阻塞队列···202

6.1 Queue 接口···202
6.2 BlockingQueue 接口··202
6.3 BlockingQueue 实现类···203
6.4 LinkedBlockingQueue 与 ArrayBlockingQueue···203
  6.4.1 阻塞队列的单锁与双锁··204
  6.4.2 ArrayBlockingQueue 并发分析··205
  6.4.3 LinkedBlockingQueue 并发分析··206

6.4.4	案例：12306 抢票	208

6.5 生产者与消费者模式 210
  6.5.1 基于管道发送与接收消息 211
  6.5.2 基于阻塞队列发送与接收消息 213
  6.5.3 案例：医院挂号 213
6.6 SynchronousQueue 217
  6.6.1 同步队列应用场景 217
  6.6.2 案例：Web 服务器处理并发请求 218
6.7 延迟阻塞队列 220
  6.7.1 案例：元素延迟出队 221
  6.7.2 项目案例：Web 服务器会话管理 222
6.8 PriorityBlockingQueue 226
  案例：按优先级执行任务 226
6.9 LinkedTransferQueue 228
6.10 LinkedBlockingDeque 229
6.11 本章习题 229

第 7 章　线程池与 AQS 231

7.1 acquire 与 release 231
7.2 性能目标 232
7.3 设计与实现 233
  7.3.1 同步状态 233
  7.3.2 阻塞 244
  7.3.3 排队 246
  7.3.4 条件队列 248
7.4 使用 AQS 249
  7.4.1 控制公平性 250
  7.4.2 同步器 251
7.5 AQS 性能 252
  7.5.1 过载 253
  7.5.2 吞吐量 254
7.6 本章习题 255

第 8 章　结束线程与线程池任务 257

8.1 stop()与 destroy() 257
8.2 状态值结束线程 258

- 8.3 shutdown()与shutdownNow() ············ 258
- 8.4 线程休眠 ············ 258
- 8.5 线程中断 ············ 258
- 8.6 Future 与 FutureTask ············ 258
  - 8.6.1 取消任务 ············ 259
  - 8.6.2 任务超时结束 ············ 263
- 8.7 项目案例：所有线程池任务暂停与重启 ············ 264
- 8.8 本章习题 ············ 267

第 9 章 Tomcat 线程池技术 ············ 268

- 9.1 自定义 ThreadPoolExecutor ············ 268
- 9.2 Tomcat 任务队列 ············ 270
- 9.3 Tomcat 任务线程 ············ 270
- 9.4 Tomcat 任务线程工厂 ············ 271
- 9.5 Tomcat 连接器与线程池 ············ 272
- 9.6 创建 Tomcat 线程池 ············ 274
- 9.7 Web 服务器异步环境 ············ 275
  - 案例：AsyncContext 调用业务方法 ············ 276
- 9.8 Web 服务器 NIO ············ 278
  - 案例：服务器 NIO 处理请求 ············ 279
- 9.9 本章习题 ············ 281

第 10 章 并发编程应用 ············ 283

- 10.1 JVM 与多线程 ············ 283
- 10.2 Servlet 与多线程 ············ 284
- 10.3 懒汉与恶汉模式 ············ 286
- 10.4 数据库 Connection 与多线程 ············ 288
  - 10.4.1 ThreadLocal 与线程私有数据 ············ 289
  - 10.4.2 ThreadLocal 存储数据库 Connection ············ 291
  - 10.4.3 ThreadLocal 实现 Connection per logic 模式 ············ 293
  - 10.4.4 ThreadLocal 实现 Connection per request 模式 ············ 294
- 10.5 高并发网站的 PageView 统计 ············ 295
- 10.6 生成唯一的订单号 ············ 296
- 10.7 浏览器并发请求限制 ············ 298

10.8　NIO 与多路复用 301
10.9　远程异步访问 302
10.10　防止缓存雪崩的 DCL 机制 305
10.11　分布式锁解决商品超卖 309

参考文献 314

第 1 章 多线程基础

进程是操作系统管理应用程序的基本单元，而线程是编写并发程序的基础。编写并发程序不是一件容易的事情，即便你已经理解并掌握了线程的基本操作也绝不简单。

1.1 进程与线程

1.1.1 操作系统与进程

早期的操作系统有 DOS、UNIX、MAC 等，直到 1995 年，微软著名的 Windows 系统逐渐占领客户端市场，而服务器的操作系统则由更加稳定的 Linux 统治。

分时操作系统是早期操作系统的主流：一台主机连接了若干个终端；每个终端有一个用户在使用，交互式地向系统提出命令请求，系统接受每个用户的命令，然后采用时间片轮转方式处理服务请求，并通过交互方式在终端上向用户显示处理结果。

分时操作系统中的每一个进程都是一个虚拟的冯·诺依曼机。它拥有一个内存空间，存储着指令和数据，根据机器语言的语义来顺序执行指令，并且通过操作系统的 I/O 原语集来实现与外部世界的交互。对于每一条指令的执行，都有一个对"下一条指令"的明确定义，并根据程序中的指令集来进行流程的控制。现在几乎所有广泛使用的编程语言都遵循这个顺序的编程模型，其中语言规范明确定义了在一个给定动作完成后，下一个动作是什么。

这种工作模式存在着一个严重问题，阻塞让这一切显得不那么合理。机器只执行一个操作，只有在前一个操作执行完成后，才能执行后继操作。这种顺序执行的模式使得应用程序在运行时独占全部计算机资源，资源利用非常低。

多任务操作系统的发展使得多个应用程序能够同时运行，程序在各自的进程中彼此独立运行，由操作系统分配资源，诸如内存、文件句柄等。进程之间可以通过信号量、共享数据、socket 等彼此通信。

进程是操作系统动态执行的基本单元，在传统的操作系统中，进程既是基本的分配单元，也是基本的执行单元。任何一个应用程序都可以看作是一个运行时的进程。进程是运行在它自己的地址空间的自包容程序。多任务操作系统可以通过周期性地将 CPU 从一个进程切换到另一个进程，来实现同时运行多个进程。

打开 Windows 任务管理器，在任务列表中你可以直观地看到当前系统正在运行的进程信息，如图 1-1 所示。

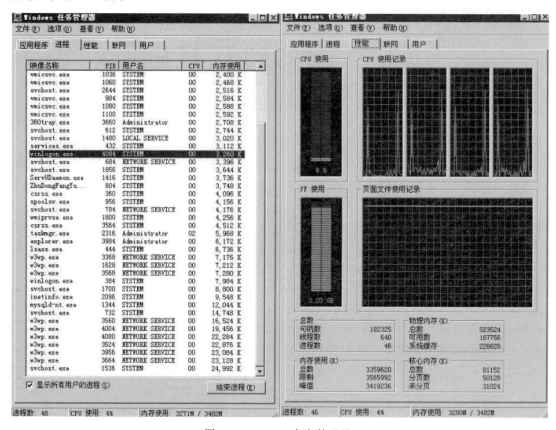

图 1-1 Windows 任务管理器

1.1.2 线程与进程的关系

进程可以看成是线程的容器，而线程又可以看成是进程中的执行路径。线程使得程序控制流的多个分支可以执行在一个进程中，它们共享这个进程范围内的所有资源。同一应用程序中的多个线程可以在多处理器系统中被同时调度。在大多数现代操作系统中把线程作为时序调度的基本单元。同一个进程中的多个线程同步或者异步执行，会共享其所属进程的内存地址空间，因此线程之间可以很容易地实现数据共享。但是如果没有同步措施，多个线程同时修改共享数据将会产生意想不到的结果。

多线程是一把双刃剑。在 Java 中，恰当地设计并使用多线程可以提高一个复杂应用的性能，降低开发成本。例如，在 Tomcat 中线程被用来提高资源的利用率和吞吐率。但是稍有不慎可能使得应用程序面临未知异常，甚至是崩溃。

Java 的多线程机制是抢占式的，这表明调度机制会周期性地中断线程，将上下文切换到另一个线程，从而为每个线程都提供 CPU 时间分片，使得每个线程都会分配到数量合理的时间去驱动它的任务。

每个进程运行时至少会有 1 个主线程运行，如 JavaSE 的应用，调用 main()函数就会启动主线程。代码示例如下：

```java
public static void main(String[] args){}
```

那么到底什么是线程呢？简单总结如下：线程（Thread）的字面意思是线路，即应用程序（进程）中的程序执行线路。Java 虚拟机允许一个应用程序中可以同时并发存在多条程序执行线路。每个线程都有一个优先级属性，优先级别高的线程，可能会被 CPU 优先执行。归根到底，线程就是应用程序（进程）在运行过程中，通过操作系统向 CPU 发起的一个任务，这个任务只能访问当前进程的内存资源。

1.2 多线程启动

线程有两种启动方式：实现 Runnable 接口；继承 Thread 类并重写 run()方法。

执行进程中的任务时才会产生线程，因此需要一种描述任务的方式，这可以由 Runnable 接口来提供。要想定义任务，需要实现 Runnable 接口并且重写 run()方法，然后再将 Runnable 的实现对象作为参数传递给 Thread 类。

调用 Thread 类的 start()方法，启动线程，向 CPU 发出请求，去执行任务。

代码示例如下：

```java
public class MyTest {
    public static void main(String[] args) {
        //创建任务
        MyRunnable mr = new MyRunnable();
        //启动线程，申请执行任务
        new Thread(mr).start();
    }
}
//实现 Runnable 接口，定义一个任务
class MyRunnable implements Runnable {
    public void run() {
        System.out.println("我的任务开始执行...");
    }
}
```

还可以采用继承 Thread 类并且重写 run()方法，然后调用 start()启动线程。

代码示例如下：

```java
public class ThreadStart {
```

```java
    public static void main(String[] args){
        MyThread mt = new MyThread();
        mt.start();
    }
}
class MyThread extends Thread {
    //重写父类的 run 方法
    public void run(){
        System.out.println("thread running...");
    }
}
```

通常情况下，实现 Runnable 接口然后启动线程是一个更好的选择，这可以提高程序的灵活性和扩展性，并且用 Runnable 接口描述任务也更容易理解。在后面的线程池调用中，也使用 Runnable 表示要执行的任务。

需要特别注意的是，执行 start()方法的顺序不代表线程启动的顺序，在下面示例的 ThreadTest 中，我们按照顺序调用了 8 个线程的 start()方法，但是线程的执行顺序并没有规律，而且每次运行的结果可能都不一样。

代码示例如下：

```java
public class ThreadTest {
    public static void main(String[] args){
        Thread t1 = new Thread(new MyThread(1));
        Thread t2 = new Thread(new MyThread(2));
        Thread t3 = new Thread(new MyThread(3));
        Thread t4 = new Thread(new MyThread(4));
        Thread t5 = new Thread(new MyThread(5));
        Thread t6 = new Thread(new MyThread(6));
        Thread t7 = new Thread(new MyThread(7));
        Thread t8 = new Thread(new MyThread(8));
        t1.start();
        t2.start();
        t3.start();
        t4.start();
        t5.start();
        t6.start();
        t7.start();
        t8.start();
    }
    static class MyThread implements Runnable {
        private int no;
        public MyThread(int no){
```

```
            this.no = no;
        }
        public void run(){
            System.out.println(no);
        }
    }
}
```

执行结果如下:

```
1
7
6
5
2
3
4
8
process finished with exit code 0
```

为什么会出现这样的运行结果呢？这主要是因为任务的执行靠 CPU，而处理器采用分片轮询方式执行任务，所有的任务都是抢占式执行模式，也就是说任务是不排序的。可以设置任务的优先级，优先级高的任务可能会优先执行（多数时候是无效的）。任务被执行前，该线程处于自旋等待状态。

1.2.1 线程标识

Thread 类用于管理线程，如设置线程优先级、设置 Daemon 属性、读取线程名字和 ID、启动线程任务、暂停线程任务、中断线程等。

为了管理线程，每个线程在启动后都会生成一个唯一的标识符，并且在其生命周期内保持不变。当线程被终止时，该线程 ID 可以被重用。而线程的名字更加直观，但是不具有唯一性。

代码示例如下:

```
public static void main(String[] args) {
    for(int i=0;i<5;i++) {
        MyRunnable mr = new MyRunnable();
        //启动线程，申请执行任务
        Thread th = new Thread(mr);
        th.start();
        System.out.println(th.getId());
        System.out.println(th.getName());
    }
```

```java
}
class MyRunnable implements Runnable {
    public void run() {
        System.out.println("我的任务开始执行...");
    }
}
```

程序运行结果如下：

```
8
我的任务开始执行...
Thread-0
9
Thread-1
我的任务开始执行...
10
Thread-2
我的任务开始执行...
11
Thread-3
我的任务开始执行...
12
Thread-4
我的任务开始执行...
```

1.2.2 Thread 与 Runnable

Runnable 接口表示线程要执行的任务。当 Runnable 中的 run()方法执行时，表示线程在激活状态，run()方法一旦执行完毕，即表示任务完成，则线程将被停止。

```java
public interface Runnable {
    void run();
}
```

Thread 类默认实现了 Runnable 接口，并且其构造方法的重载形式允许传入 Runnable 接口对象作为任务。

Thread 类的源代码片段如下：

```java
public class Thread implements Runnable {
    private Runnable target;
    public void run() {
        if (target != null) {
            target.run();
        }
```

```
    }
}
```

通过 Thread 类的源代码可以发现，线程的两种启动方式，其本质都是实现 Thread 类中的 run()方法。而实现 Runnable 接口，然后传递给 Thread 类的方式，比 Thread 子类重写 run()方法更加灵活。

1.2.3　run()与 start()

Thread 类中的 run()方法与 start()方法很容易混淆，我们对比分析一下。

```java
public class Thread implements Runnable{
    public void run() {
    }
    public synchronized void start() {
    }
}
```

调用 Thread 对象的 start()方法，使线程对象开始执行任务，这会触发 Java 虚拟机调用当前线程对象的 run()方法。调用 start()方法后，将导致两个线程并发运行，一个是调用 start()方法的当前线程，另外一个是执行 run()方法的线程。

如果重复调用 start()方法，这是一个非法操作，它不会产生更多的线程，反而会导致 IllegalThreadStateException 异常。

代码测试如下：

```java
public class MyTest {
    public static void main(String[] args) {
        //创建任务
        MyRunnable mr = new MyRunnable();
        //创建线程对象，任务还未执行
        Thread th = new Thread(mr);
        //启动线程，申请执行任务
        th.start();
        System.out.println("主线程：" + Thread.currentThread().getId());
    }
}
class MyRunnable implements Runnable {
    public void run() {
        System.out.println(Thread.currentThread().getId()+"-我的任务开始执行...");
    }
}
```

测试结果如下：

主线程：1
8-我的任务开始执行...

重复调用 start()方法，将会引发异常，测试代码如下：

```java
public static void main(String[] args) {
    //创建任务
    MyRunnable mr = new MyRunnable();
    //创建线程对象，任务还未执行
    Thread th = new Thread(mr);
    //启动线程，申请执行任务
    th.start();
    //再次调用 start()会引发异常
    th.start();
    System.out.println("主线程: " + Thread.currentThread().getId());
}
```

测试结果如下：

```
8-我的任务开始执行...
Exception in thread "main" java.lang.IllegalThreadStateException
at java.lang.Thread.start(Thread.java:705)
at com.icss.ui.MyTest.main(MyTest.java:12)
```

这里要特别注意的是，Thread 对象调用 start()方法之后，触发了 JVM 底层调用 run()方法。如果我们主动调用 Thread 对象的 run()方法，并不能启动一个新线程。

代码测试如下：

```java
public static void main(String[] args) {
    //创建任务
    MyRunnable mr = new MyRunnable();
    //创建线程对象，任务还未执行
    Thread th = new Thread(mr);
    //线程对象直接调用 run()方法，不会启动新线程
    th.run();
    System.out.println("主线程: " + Thread.currentThread().getId());
}
```

测试结果如下，执行 run()方法的线程与主线程 ID 一样，表明它们是同一个线程：

```
1-我的任务开始执行...
主线程：1
```

1.2.4　Thread 源码分析

创建 Thread 类实例，首先会执行 registerNatives()方法，它在静态代码块中加载。线程的启动、运行、生命期管理和调度等都高度依赖于操作系统，Java 本身并不具备与底层操作系统交互的能力。因此线程的底层操作都使用了 native 方法，registerNatives()就是用 C 语言编写的底层线程注册方法。

```java
public class Thread implements Runnable {
    private static native void registerNatives();
    static {
        registerNatives();
    }
}
```

无论通过 Thread 类的哪种构造方法去创建线程，都需要首先调用 init()方法，初始化线程环境，源码如下：

```java
public class Thread {
    public Thread() {
        init(null, null, "Thread-" + nextThreadNum(), 0);
    }
    public Thread(Runnable target) {
        init(null, target, "Thread-" + nextThreadNum(), 0);
    }
}
private void init(ThreadGroup g, Runnable target, String name,
                  long stackSize, AccessControlContext acc) {
    if (name == null) {
        throw new NullPointerException("name cannot be null");
    }
    this.name = name;
    Thread parent = currentThread();
    SecurityManager security = System.getSecurityManager();
    if (g == null) {
        if (security != null) {
            g = security.getThreadGroup();
        }
        if (g == null) {
            g = parent.getThreadGroup();
        }
    }
    g.checkAccess();
```

```java
        if (security != null) {
            if (isCCLOverridden(getClass())) {
                security.checkPermission(SUBCLASS_IMPLEMENTATION_PERMISSION);
            }
        }
        g.addUnstarted();
        this.group = g;
        this.daemon = parent.isDaemon();
        this.priority = parent.getPriority();
        if (security == null || isCCLOverridden(parent.getClass()))
            this.contextClassLoader = parent.getContextClassLoader();
        else
            this.contextClassLoader = parent.contextClassLoader;
        this.inheritedAccessControlContext =
                acc != null ? acc : AccessController.getContext();
        this.target = target;
        setPriority(priority);
        if (parent.inheritableThreadLocals != null)
            this.inheritableThreadLocals =
                ThreadLocal.createInheritedMap(parent.inheritableThreadLocals);
        this.stackSize = stackSize;
        tid = nextThreadID();
}
```

在 init() 方法中，做了如下操作：

（1）设置线程名称。

（2）将新线程的父线程设置为当前线程。

（3）获取系统的安全管理 SecurityManager，并获得线程组。SecurityManager 在 Java 中被用来检查应用程序是否能访问一些受限资源，如文件、套接字（socket）等。它可以用在那些具有高安全性要求的应用程序中。

（4）获取线程组的权限检查。

（5）在线程组中增加未启动的线程数量。

（6）设置新线程的属性，包括守护线程属性（默认继承父线程）、优先级（默认继承父线程）、堆栈大小（如果为 0，则默认由 JVM 分配）、线程组、线程安全控制上下文（一种 Java 安全模式，设置访问控制权限）等。

接下来我们看一下 Thread 类中的 start() 方法，源码如下：

```java
public synchronized void start() {
    if (threadStatus != 0)
        throw new IllegalThreadStateException();
    group.add(this);
```

```java
    boolean started = false;
    try {
        start0();
        started = true;
    } finally {
        try {
            if (!started) {
                group.threadStartFailed(this);
            }
        } catch (Throwable ignore) {}
    }
}
```

它做了如下操作：
（1）在线程组中减少未启动的线程数量。
（2）调用底层的 native 方法 start()进行线程启动。
（3）最终由底层 native 方法调用 run()执行。
（4）如果启动失败，从线程组中移除该线程，并且增加未启动线程数量。

1.3 线程状态

线程对象在不同的运行时期存在着不同的状态，在 Thread 类中通过一个内部枚举类 State 保存状态信息，了解线程状态对于并发编程非常重要。

```java
public class Thread {
    public enum State {
        NEW,
        RUNNABLE,
        BLOCKED,
        WAITING,
        TIMED_WAITING,
        TERMINATED;
    }
}
```

Java 中的线程存在 6 种状态，分别是 NEW、RUNNABLE、BLOCKED、WAITING、TIMED_WAITING、TERMINATED。我们可以通过 Thread 类中的 Thread.getState()方法获取线程在某个时期的线程状态。在给定的时间点，线程只能处于一种状态。

1.3.1 NEW 状态

NEW 代表着线程新建状态，一个已创建但是未起动（start）的线程处于 NEW 状态。

测试代码如下：

```java
public static void main(String[] args) {
  Thread th = new Thread(new Runnable() {
    public void run() {
      System.out.println(Thread.currentThread().getId()
                          + "-我的任务开始执行...");
    }
  });
  System.out.println(th.getName() + ":状态: " + th.getState());
  System.out.println("主线程: " + Thread.currentThread().getId());
}
```

测试结果如下：

```
Thread-0:状态: NEW
主线程: 1
```

1.3.2　RUNNABLE 状态

RUNNABLE 状态表示一个线程正在 Java 虚拟机中运行，但是这个线程是否获得了处理器分配资源并不确定。调用 Thread 的 start()方法后，线程从 NEW 状态切换到了 RUNNABLE 状态。

代码测试如下：

```java
public static void main(String[] args) {
  Thread th = new Thread(new Runnable() {
    public void run() {
      System.out.println(Thread.currentThread().getName() +
                ":状态: " + Thread.currentThread().getState());
      System.out.println(Thread.currentThread().getName() +
                "-我的任务开始执行...");
      System.out.println(Thread.currentThread().getName() +
                ":状态: " + Thread.currentThread().getState());
    }
  });
  System.out.println(th.getName() + ":状态: " + th.getState());
  th.start();
  System.out.println("主线程: " + Thread.currentThread().getName() +
                ":状态: " + Thread.currentThread().getState());
}
```

测试结果如下：

```
Thread-0:状态: NEW
主线程：main:状态：RUNNABLE
Thread-0:状态: RUNNABLE
Thread-0-我的任务开始执行...
Thread-0:状态: RUNNABLE
```

1.3.3 BLOCKED 状态

BLOCKED 为阻塞状态，表示当前线程正在阻塞等待获得监视器锁。当一个线程要访问被其他线程 synchronized 锁定的资源时，当前线程需要阻塞等待。

代码测试如下，在主函数中分别启动了两个线程，它们都需要获得 object 对象的监视器锁后执行任务。第一个线程启动后，首先获得了 object 监视器锁。由于在 run()方法中使用了死循环 while(true)，因此第一个线程获得了 object 监视锁后不会释放，这导致第二个线程长期处于阻塞等待状态。

第一个线程的状态经历了 NEW→RUNNABLE 状态的变化；第二个线程的状态经历了 NEW→RUNNABLE→BLOCKED 等几个状态的变化。

```java
public static void main(String[] args) {
    Object object = new Object();
    Thread th = new Thread(new Runnable() {
        public void run() {
            synchronized(object) {
                while(true) {
                }
            }
        }
    });
    Thread th2 = new Thread(new Runnable() {
        public void run() {
            synchronized(object) {
                while(true) {
                }
            }
        }
    });
    System.out.println(th.getName() + ":状态: " + th.getState());
    System.out.println(th2.getName() + ":状态: " + th2.getState());
    th.start();
    System.out.println(th.getName() + ":状态: " + th.getState());
    for(int i=0;i<3000000;i++) {
        //适当延迟后，启动第二个线程
    }
```

```java
        th2.start();
        System.out.println(th2.getName() + ":状态: " + th2.getState());
        for(int i=0;i<3000000;i++) {
            //适当延迟后,重新获取线程状态
        }
        System.out.println(th.getName() + ":状态: " + th.getState());
        System.out.println(th2.getName() + ":状态: " + th2.getState());
        System.out.println("主线程: " + Thread.currentThread().getName()
                + ":状态: " + Thread.currentThread().getState());
    }
```

测试结果如下:

```
Thread-0:状态: NEW
Thread-1:状态: NEW
Thread-0:状态: RUNNABLE
Thread-1:状态: RUNNABLE
Thread-0:状态: RUNNABLE
Thread-1:状态: BLOCKED
主线程: main:状态: RUNNABLE
```

1.3.4 WAITING 状态

WAITING 表示线程处于等待状态。

在当前线程中调用如下方法之一时,会使当前线程进入等待状态:

Object 类的 wait()方法(没有超时设置);

Thread 类的 join()方法(没有超时设置);

LockSupport 类的 park()方法。

处于等待状态的线程,正在等待另外一个线程去完成某个特殊操作。例如,在某个线程中调用了 Object 对象的 wait()方法,它会进入等待状态,等待 Object 对象调用 notify()或 notifyAll()方法。一个线程对象调用了 join()方法,则会等待指定的线程终止任务。

代码测试一如下,注意调用 object 对象的 wait()方法时,必须要先用 synchronized 锁定 object 对象。当线程 1 进入 WAITING 状态后,后续代码不再执行。直到 object 对象调用了 notify()方法,线程 1 才继续运行。

```java
public static void main(String[] args) {
    Object object = new Object();
    Thread t1 = new Thread(new Runnable() {
        public void run() {
            for(int i=0;i<10;i++) {
                System.out.println(Thread.currentThread().getName()+",i="+i);
                if(i==5) {
```

```java
            synchronized(object) {
                try {
                    System.out.println(Thread.currentThread().getName()
                                    + "开始等待...");
                    object.wait();
                } catch (Exception e) {
                    e.printStackTrace();
                }
            }
        } });
        Thread t2 = new Thread(new Runnable() {
            public void run() {
                System.out.println(Thread.currentThread().getName() + " running...");
                synchronized(object) {
                    try {
                        System.out.println(Thread.currentThread().getName()
                                    + ",发送notify通知...");
                        object.notify();
                    } catch (Exception e) {
                        e.printStackTrace();
                    }
                }
        } });
        //启动第一个线程
        t1.start();
        double d = 0;
        for(int i=0;i<10000000;i++) {
            //等待第一个线程调用wait()方法
            d+=(Math.PI + Math.E)/(double)i;
        }
        //在主线程读取第一个线程的状态
        System.out.println(t1.getName()+ "状态: " + t1.getState());
        //启动第二个线程,发送notify()通知
        t2.start();
}
```

测试结果如下:

```
Thread-0,i=0
Thread-0,i=1
Thread-0,i=2
Thread-0,i=3
Thread-0,i=4
Thread-0,i=5
Thread-0 开始等待...
```

```
Thread-0 状态：WAITING
Thread-1 running...
Thread-1，发送notify通知...
Thread-0,i=6
Thread-0,i=7
Thread-0,i=8
Thread-0,i=9
```

代码测试二如下，注意调用 wait()/notify()方法前需要先锁定 object 对象，而调用 park()和 unpark()方法前无须锁定对象。

```java
public static void main(String[] args) {
    Thread t1 = new Thread(new Runnable() {
        public void run() {
            for(int i=0;i<10;i++) {
                System.out.println(Thread.currentThread().getName()+",i="+i);
                if(i==5) {
                    System.out.println(Thread.currentThread().getName()
                            + "开始等待...");
                    LockSupport.park();
                }
            }
        }
    });
    Thread t2 = new Thread(new Runnable() {
        public void run() {
            System.out.println(Thread.currentThread().getName() + " running...");
            System.out.println(Thread.currentThread().getName() + ",发送解锁通知...");
            LockSupport.unpark(t1);
        }
    });
    t1.start();
    double d = 0;
    for(int i=0;i<10000000;i++) {
        //等待第一个线程调用park()方法
        d+=(Math.PI + Math.E)/(double)i;
    }
    System.out.println(t1.getName()+ "状态：" + t1.getState());
    t2.start();
}
```

测试结果如下：

```
Thread-0,i=0
Thread-0,i=1
Thread-0,i=2
Thread-0,i=3
Thread-0,i=4
Thread-0,i=5
Thread-0 开始等待...
Thread-0 状态：WAITING
Thread-1 running...
Thread-1,发送解锁通知...
Thread-0,i=6
Thread-0,i=7
Thread-0,i=8
Thread-0,i=9
```

代码测试三如下，在 t2 线程中调用 t1 线程的 join()方法，导致 t2 线程进入 WAITING 状态，直到 t1 线程全部执行完毕，t2 线程才继续执行。

```java
public static void delay() {
    double d = 0;
    for(int i=0;i<10000000;i++) {
        //浮点运算时间较长，可以起到延时作用
        d+=(Math.PI + Math.E)/(double)i;
    }
}
static Thread t1=null;
static Thread t2=null;
public static void main(String[] args) {
    t1 = new Thread(new Runnable() {
    public void run() {
        for(int i=0;i<10;i++) {
            System.out.println(Thread.currentThread().getName()
                            + ",i=" +i);
            delay();
        }
        System.out.println("t2 线程状态:" + t2.getState());
        System.out.println("t1 运行结束...");
    }
    });
    t2 = new Thread(new Runnable() {
      public void run() {
        for(int k=0;k<10;k++) {
            System.out.println(Thread.currentThread().getName()
```

```java
                                      + ",k=" +k);
                    delay();
                    if(k==3) {
                        try {
                            System.out.println("t1 加入 t2...");
                            t1.join();   //t1 加入 t2 线程，导致 t2 线程进入 WAITING 状态
                        } catch (Exception e) {
                            e.printStackTrace();
                        }
                    }
                }
            }
        });
        t1.start();
        t2.start();
    }
```

测试结果如下：

```
Thread-1,k=0
Thread-0,i=0
Thread-0,i=1
Thread-1,k=1
Thread-1,k=2
Thread-0,i=2
Thread-1,k=3
Thread-0,i=3
t1 加入 t2....
Thread-0,i=4
Thread-0,i=5
Thread-0,i=6
Thread-0,i=7
Thread-0,i=8
Thread-0,i=9
t2 线程状态:WAITING
t1 运行结束...
Thread-1,k=4
Thread-1,k=5
Thread-1,k=6
Thread-1,k=7
Thread-1,k=8
Thread-1,k=9
```

1.3.5 TIMED_WAITING 状态

TIMED_WAITING 表示线程处于定时等待状态。

在当前线程中调用如下方法之一时，使当前线程进入定时等待状态：

Object 类的 wait()方法（有超时设置）；

Thread 类的 join()方法（有超时设置）；

Thread 类的 sleep()方法（有超时设置）；

LockSupport 类的 parkNanos ()方法；

LockSupport 类的 parkUntil()方法。

代码测试一如下，调用 object.wait(3000)，在指定时间内没有调用 object 对象的 notify()或 notifyAll()，就会触发超时等待结束，当前线程重新进入 RUNNABLE 状态。

```java
public static void main(String[] args) {
    Object object = new Object();
    Thread t1 = new Thread(new Runnable() {
     public void run() {
        for(int i=0;i<10;i++) {
            System.out.println(Thread.currentThread().getName()
                            + ",i=" +i);
            if(i==5) {
                synchronized(object) {
                    try {
                        System.out.println(Thread.currentThread().getName()
                                        + "开始等待...");
                        object.wait(3000);
                    } catch (Exception e) {
                        e.printStackTrace();
                    }
                }
    } } } });
    t1.start();
    try {
        //延迟1秒，等待t1进入等待状态
         Thread.sleep(1000);
    } catch (Exception e) {
    }
    System.out.println(t1.getName()+ "状态: " + t1.getState());
}
```

测试结果如下：

Thread-0,i=0

```
Thread-0,i=1
Thread-0,i=2
Thread-0,i=3
Thread-0,i=4
Thread-0,i=5
Thread-0 开始等待...
Thread-0 状态：TIMED_WAITING
Thread-0,i=6
Thread-0,i=7
Thread-0,i=8
Thread-0,i=9
```

代码测试二如下，调用 Thread.*sleep*(3000)，当前线程也会进入定时等待状态。

```java
public static void main(String[] args) {
    Thread t1 = new Thread(new Runnable() {
        public void run() {
            for (int i = 0; i < 10; i++) {
                System.out.println(Thread.currentThread().getName() + ",i=" + i);
                if (i == 5) {
                    try {
                        System.out.println(Thread.currentThread().getName()
                                + "开始等待...");
                        Thread.sleep(3000);
                    } catch (Exception e) {
                    }
                }
            }
        } });
    t1.start();
    try {
        //延迟1秒，等待 t1 进入等待状态
        Thread.sleep(1000);
    } catch (Exception e) {
    }
    System.out.println(t1.getName() + "状态：" + t1.getState());
}
```

测试结果如下：

```
Thread-0,i=0
Thread-0,i=1
Thread-0,i=2
Thread-0,i=3
Thread-0,i=4
```

```
Thread-0,i=5
Thread-0 开始等待...
Thread-0 状态：TIMED_WAITING
Thread-0,i=6
Thread-0,i=7
Thread-0,i=8
Thread-0,i=9
```

1.3.6　WAITING 与 BLOCKED 的区别

WAITING、TIMED_WAITING、BLOCKED 这几个线程状态，都会使当前线程处于停顿状态，因此容易混淆。下面简单总结一下这些状态之间的区别：

（1）Thread.sleep()不会释放占有的对象锁，因此会持续占用 CPU。

（2）Object.wait()会释放占有的对象锁，不会占用 CPU。

（3）BLOCKED 使当前线程进入阻塞后，为了抢占对象监视器锁，一般操作系统都会给这个线程持续的 CPU 使用权。

（4）LockSupport.park()底层调用 UNSAFE.park()方法实现，它没有使用对象监视器锁，不会占用 CPU。

1.3.7　TERMINATED 状态

TERMINATED 表示线程为完结状态。当线程完成其 run()方法中的任务，或者因为某些未知的异常而强制中断时，线程状态变为 TERMINATED。

代码测试如下：

```java
public static void main(String[] args) {
    Thread t1 = new Thread(new Runnable() {
        public void run() {
            double d = 0;
            for(int i=0;i<1000;i++) {
                d+=(Math.PI + Math.E)/(double)i;
            }
        }
    });
    t1.start();
    try {
        Thread.sleep(2000);
    } catch (Exception e) {
    }
    System.out.println("t1 线程状态：" + t1.getState());
}
```

1.3.8 线程状态转换

前面我们学习了 Java 线程的 6 种状态，接下来通过图 1-2 对线程状态转换进行汇总。在图 1-2 中，把线程 RUNNABLE 状态细分为两种：runnable（准备就绪）和 running（运行中）。runnable 表示线程刚刚被 JVM 启动，还没有获得 CPU 的使用权；running 表示线程获得了 CPU 的使用权，正在运行。

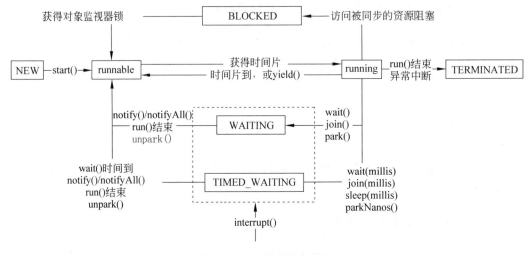

图 1-2　Java 线程状态转换

1.4　sleep()与 yield()

1.4.1　线程休眠 sleep()

Thread 类的 sleep()方法，使当前正在执行的线程以指定的毫秒数暂时停止执行，具体停止时间取决于系统定时器和调度程序的精度和准确性。当前线程状态由 RUNNABLE 切换到 TIMED_WAITING。调用 sleep()方法不会使线程丢失任何监视器所有权，因此当前线程仍然占用 CPU 分片。

```
public class Thread {
    public static native void sleep(long millis)
            throws InterruptedException;
}
```

调用 sleep()方法可能会抛出 InterruptedException 异常，它应该在 run()方法中被捕获，因为异常无法传递到其他线程，如主线程就无法捕获子线程抛出的异常。

Java SE5 引入了更加显式的 sleep()版本，作为 TimeUit 类的一部分。例如，TimeUnit.

MILLISECONDS.sleep(1000)等价于 Thread.sleep(1000),表示休眠 1 秒。TimeUnit 类提供了更好的可读性。

代码测试如下：

```java
public class MyTest {
    public static void main(String[] args){
        Task t = new Task();
        new Thread(t).start();
    }}
class Task implements Runnable {
  public void run(){
    try {
        long beginTime = System.currentTimeMillis();
      System.out.println("每隔1秒输出[0,10)区间的整数,开始...");
      for(int i = 0; i < 10; i++) {
          TimeUnit.SECONDS.sleep(1);
          System.out.println("数字: " + i );
      }
      System.out.println("每隔1秒输出[0,10)区间的整数,结束...");
      long endTime = System.currentTimeMillis();
      System.out.println("共耗时: " + (endTime-beginTime)/1000 + "秒.");
    } catch (InterruptedException e) {
      e.printStackTrace();
    }
}}
```

测试结果如下：

```
每隔 1 秒输出[0,10)之间的整数，开始...
数字：0
数字：1
数字：2
数字：3
数字：4
数字：5
数字：6
数字：7
数字：8
数字：9
每隔 1 秒输出[0,10)之间的整数，结束...
共耗时：10 秒.
```

在上面的示例中，使用独立线程输出[0，10)区间的整数，每输出一个数字，线程通过

TimeUnit.SECONDS.sleep(1)语句休眠 1 秒。

1.4.2 线程让步 yield()

Thread 类的 yield()方法对线程调度器发出一个暗示,即当前线程愿意让出正在使用的处理器。调度程序可以响应暗示请求,也可以自由地忽略这个提示。如图 1-2 所示,线程调用 yield()方法后,可能从 running 状态转为 runnable 状态。

```java
public class Thread {
    public static native void yield();
}
```

需要强调的是:yield()仅仅是一个暗示,没有任何机制保证它一定会被采纳。线程调度器是 Java 线程机制的底层对象,可以把 CPU 的使用权从一个线程转移到另外一个线程。如果你的计算机是多核处理器,那么分配线程到不同的处理器执行任务要依赖线程调度器。

代码测试如下:

(1) 新建任务类 ListOff,用于倒计时显示,使用 sleep()作为线程延时。

```java
public class ListOff implements Runnable{
    private int countDown = 5;
    public void run() {
        while(countDown-- > 0) {
            String info = Thread.currentThread().getId() + "#" +countDown;
            System.out.println(info);
            try {
                TimeUnit.MILLISECONDS.sleep(100);
            } catch (Exception e) {
            }
        }
    }
}
```

(2) 创建一个倒计时器,两个线程同时使用。参见运行结果可知,每次的运行结果都不同,而且都不正确。其关键原因是 countDown 唯一,两个线程可能同时访问这块内存,后面我们可以通过加锁的方式解决这个问题。

```java
public static void main(String[] args){
    ListOff lf = new ListOff(); //创建一个倒计时器
    new Thread(lf).start();
    new Thread(lf).start();
}
```

运行结果如下:

8#4   9#3   8#2   9#2   8#0

第二次运行结果如下:

8#3   9#3   8#1   9#1   9#-1

第三次运行结果:

9#3   8#3   9#2   8#1   9#0

(3)把 sleep()代码修改为 yield(),三次测试的结果输出都是正确的。而且 yield()方法起到了线程让步的效果(此处没有使用锁,不能保证每次的运行结果都正确)。

```java
public class ListOff implements Runnable{
    private int countDown = 5;
    public void run() {
        while(countDown-->0) {
            String info = Thread.currentThread().getId() + "#" +countDown;
            System.out.println(info);
            Thread.yield();   //让步暗示
        }
    }
}
```

运行结果如下:

8#4   9#3   8#2   9#1   8#0

第二次运行结果如下:

8#4   9#3   9#1   9#0   8#2

第三次运行结果:

9#4   9#2   8#3   9#1   8#0

1.5 线程优先级

每个线程都有优先级。具有较高优先级的线程可能优先获得 CPU 的使用权。创建一个新的 Thread 对象时,新线程的优先级默认与创建线程的优先级一致。

JDK 中实际上存在着 10 个优先级,但是这与大多数操作系统不能建立很好的映射关系。比如 Windows 有 7 个线程优先级设置,而 Sun 的 Solaris 只有两个线程优先级,因此在 Java 中一般只使用下面的三种优先级设置。

```java
public class Thread {
```

```java
    public final static int MIN_PRIORITY = 1;
    public final static int NORM_PRIORITY = 5;
    public final static int MAX_PRIORITY = 10;
}
```

我们可以通过 Thread 类中 setPriority()方法对线程的优先级进行设置，参考 JDK 的源码如下：

```java
public final void setPriority(int newPriority) {
    ThreadGroup g;
    checkAccess();
    if (newPriority > MAX_PRIORITY || newPriority < MIN_PRIORITY) {
        throw new IllegalArgumentException();
    }
    if((g = getThreadGroup()) != null) {
        if (newPriority > g.getMaxPriority()) {
            newPriority = g.getMaxPriority();
        }
        setPriority0(priority = newPriority);
    }
}
```

这里面需要注意的是，如果设置的线程优先级小于 1（MIN_PRIORITY）或者大于 10（MAX_PRIORITY）都将抛出 IllegalArgumentException 异常。

不应该过分依赖于线程优先级的设置，理论上线程优先级高的会优先执行，但实际情况可能并不明确。例如，线程调度机制还没有来得及介入时，线程可能就已经执行完了。所以优先级具有一定的"随机性"。

1.5.1　线程优先级与资源竞争

具有较高优先级的线程会优先得到调度系统资源分配。也就是说优先级高的线程可以优先竞争共享资源。但线程的优先级调度和底层操作系统有密切的关系，在各个平台上表现不一并且无法精准控制。因此在要求严格的场合，需要开发者在应用层解决线程调度问题。

当调用 Thread.yield()方法时，会给线程调度器一个暗示，即优先级高的其他线程或相同优先级的其他线程，都可以优先获得 CPU 分片。

1.5.2　案例：大型浮点运算测试

案例场景描述：创建 6 个线程，每个线程都计算了足够量级的浮点运算（浮点运算比较费时），目的是让线程调度机制来得及介入。其中将 1 个线程的优先级设置为最高（MAX_PRIORITY=10），1 个线程的优先级设置为默认（NORM_PRIORITY=5），其余 4 个线程优先级设置为最低（MIN_PRIORITY=1）。为了保证计算的精度，在代码中使用了 BigDecimal

对象。

代码示例如下：

```java
public class FloatArithmetic implements Runnable {
    private int priority;
    public FloatArithmetic(int priority) {
        this.priority = priority;
    }
    public void run() {
        BigDecimal value = new BigDecimal("0");
        //按照参数传递值修改当前线程优先级
        Thread.currentThread().setPriority(priority);
        BigDecimal pi = new BigDecimal(Math.PI);
        BigDecimal e = new BigDecimal(Math.E);
        //足够耗时的计算，使任务调度可以反应
        for (int i = 0; i < 3000; i++) {
            for (int j = 0; j < 3000; j++) {
                value = value.add(pi.add(e).divide(pi,4));
            }
        }
        Thread self = Thread.currentThread();
        System.out.println("线程编号为" + self.getId() + ",优先级为"
                + self.getPriority() + ",计算结果为" + value.doubleValue());
    }
    public static void main(String[] args) {
        new Thread(new FloatArithmetic(Thread.MIN_PRIORITY)).start();
        new Thread(new FloatArithmetic(Thread.MIN_PRIORITY)).start();
        new Thread(new FloatArithmetic(Thread.MIN_PRIORITY)).start();
        new Thread(new FloatArithmetic(Thread.MIN_PRIORITY)).start();
        new Thread(new FloatArithmetic(Thread.NORM_PRIORITY)).start();
        new Thread(new FloatArithmetic(Thread.MAX_PRIORITY)).start();
    }
}
```

运行结果如下所示，优先级高的线程每次都会先完成计算：

线程编号为13，优先级为10，计算结果为1.6787303814890385E7
线程编号为12，优先级为5，计算结果为1.6787303814890385E7
线程编号为10，优先级为1，计算结果为1.6787303814890385E7
线程编号为11，优先级为1，计算结果为1.6787303814890385E7
线程编号为8，优先级为1，计算结果为1.6787303814890385E7
线程编号为9，优先级为1，计算结果为1.6787303814890385E7

1.5.3 案例：多线程售票

本案例使用多线程来模拟窗口售票，线程同步等相关知识后面会详细说明。
操作步骤如下：
（1）定义任务类 TicketTask，在 synchronized 同步块中需要适当延时，让高级别的线程有机会抢入。

```java
public class TicketTask implements Runnable{
    private Integer ticket = 30;
    public void run() {
        while(this.ticket>0) {
            synchronized (this) {
                if(ticket>0){
                    System.out.println("窗口" + Thread.currentThread().getId()
                                        + "售出: " + ticket);
                    ticket--;
                    try {
                        Thread.sleep(100);
                    } catch (Exception e) {
                    }
                }
            }
        }
    }}}
```

（2）在主函数中创建三个线程，分别设置不同的优先级。观察是否优先级高的线程能够获得更多的运行机会。

```java
public static void main(String[] args) {
    TicketTask task = new TicketTask();
    Thread t1 = new Thread(task);
    t1.setPriority(Thread.MIN_PRIORITY);
    Thread t2 = new Thread(task);
    Thread t3 = new Thread(task);
    t3.setPriority(Thread.MAX_PRIORITY);
    t1.start();
    t2.start();
    t3.start();
    System.out.println("t1:" + t1.getId());
    System.out.println("t2:" + t2.getId());
    System.out.println("t3:" + t3.getId());
}
```

程序运行结果如下（每次运行都会有差别）：通过反复运行可以得到结论，线程 t3 的优先级最高，虽然是最后启动的，但是 t3 每次都会获得最多的运行机会，即高优先级的线程，CPU 会优先照顾。

```
t1:8
t2:9
t3:10
窗口 8 售出：30
窗口 8 售出：29
窗口 10 售出：28
窗口 10 售出：27
窗口 10 售出：26
窗口 10 售出：25
窗口 10 售出：24
窗口 10 售出：23
窗口 10 售出：22
窗口 10 售出：21
窗口 10 售出：20
窗口 10 售出：19
窗口 10 售出：18
窗口 10 售出：17
窗口 10 售出：16
窗口 10 售出：15
窗口 10 售出：14
窗口 10 售出：13
窗口 10 售出：12
窗口 10 售出：11
窗口 10 售出：10
窗口 10 售出：9
窗口 10 售出：8
窗口 10 售出：7
窗口 10 售出：6
窗口 10 售出：5
窗口 10 售出：4
窗口 10 售出：3
窗口 10 售出：2
窗口 10 售出：1
```

1.6 守护线程

1.6.1 守护线程的概念

在 Java 线程中有两种线程，一种是用户线程，另一种是守护线程（Daemon）。

所谓守护线程，是指在程序运行的时候在后台提供一种通用服务的线程。比如，垃圾回收线程就是一个很称职的守护者（当一个对象不再被引用的时候，内存回收它占领的空间，以便空间被后来的新对象使用）。

Daemon 线程与用户线程在使用时没有任何区别，唯一的不同是：当所有用户线程结束时，程序也会终止，Java 虚拟机不管是否存在守护线程，都会退出。

调用 Thread 对象的 setDaemon()方法，可以把用户线程标记为守护者。调用 isDaemon()方法可以判断线程是否是一个守护线程。

代码测试如下：t1 为 daemon 线程，t2 为用户线程。当主线程和 t2 线程都完成任务后，t1 会被 JVM 自动结束。

```java
public static void main(String[] args) {
    Thread t1 = new Thread(new Runnable() {
        public void run() {
            while(true) {
                System.out.println(Thread.currentThread().getId() + ",run...");
                try {
                    Thread.sleep(1000);
                } catch (Exception e) {
                }
            }
        }
    });
    t1.setDaemon(true);    //设置为守护线程
    t1.start();
    Thread t2 = new Thread(new Runnable() {
        public void run() {
            for(int i=0;i<5;i++) {
                System.out.println(Thread.currentThread().getId() + ",i=" + i);
                try {
                    Thread.sleep(1000);
                } catch (Exception e) {
                }
            }
        }
```

```
        });
        t2.start();
        System.err.println(Thread.currentThread().getName() + "-主线程结束...");
}
```

测试结果如下：

```
8,run...
9,i=0
main-主线程结束...
8,run...
9,i=1
8,run...
9,i=2
8,run...
9,i=3
8,run...
9,i=4
8,run...
```

在使用守护线程时需要注意以下几点：

（1）setDaemon()方法必须在 start()方法之前设置，否则会抛出一个 IllegalThreadState-Exception 异常。不能把正在运行的常规线程设置为守护线程。

（2）在守护线程 Daemon 中产生的新线程也是守护线程，存在着继承性。

（3）守护线程应该永远不去访问固有资源，如文件、数据库，因为它会在任何时候甚至在一个操作的中间发生中断。

（4）守护线程通常都使用 while(true)的死循环来持续执行任务。

1.6.2 案例：清道夫与工作者

案例场景描述：设置守护线程判断每位员工是否可以下班。要求：员工工作时间必须大于或等于 8 小时才能够下班，否则守护进程不能同意员工下班。

代码示例如下：

（1）使用随机数模拟工人的上下班时间，工人能否下班由守护线程决定。

```java
class Worker extends Thread {
    //上班打卡时间
    private Date beginTime;
    //下班打卡时间
    private Date endTime;
    //下班状态
    private boolean isStop;
    public Worker(Date beginTime) {
```

```java
        this.beginTime = beginTime;
    }
    public void run() {
        Thread self = Thread.currentThread();
        SimpleDateFormat sdf = new SimpleDateFormat("HH:mm:ss");
        String bStr = sdf.format(beginTime);
        System.out.println(self.getId() + "号员工" + bStr + "打卡上班");
        Calendar cal = Calendar.getInstance();
        cal.setTime(beginTime);
        Random random = new Random();
        while (true) {
            try {
                TimeUnit.SECONDS.sleep(3);
            } catch (Exception e) {
            }
            if(isStop) {
                String eStr = sdf.format(endTime);
                System.out.println(self.getId()+"号员工"+eStr+"下班------");
                break;
            }else{
                int hour = random.nextInt(5);
                cal.set(Calendar.HOUR_OF_DAY, cal.get(Calendar.HOUR_OF_
                    DAY) + hour);
                endTime = cal.getTime();
            }
        }
    }
    public void setStop(boolean stop) {
        isStop = stop;
    }
    public boolean getStop() {
        return isStop;
    }
    public Date getEndTime() {
        return endTime;
    }
    public double getWorkerLongTime() {
        return (endTime.getTime() - beginTime.getTime()) / (1000 * 60 * 60);
    }
}
```

(2) 定义一个守护线程，定时监督工人的下班状态。

```java
class Sweeper extends Thread {
```

```java
    private List<Worker> workers = new ArrayList<Worker>();
    public Sweeper() {
        this.setDaemon(true);    //守护线程
    }
    public void run() {
        while (true) {
            for (int i = 0; i < workers.size(); i++) {
                Worker worker = workers.get(i);
                if (worker.getEndTime() != null && !worker.getStop()) {
                    double longTime = worker.getWorkerLongTime();
                    if (longTime >= 8) {
                        worker.setStop(true);
                        System.out.println(worker.getId() + "号员工超过 8 小
                                            时,可以下班!");
                    }else{
                        System.out.println(worker.getId() +
                                    "号员工不足 8 小时,不能下班...");
                    }
                }
                try{
                    TimeUnit.SECONDS.sleep(1);
                }catch(Exception e){
                }
            }
        }
    }
    public void addWorker(Worker worker){
        this.workers.add(worker);
    }
}
```

(3) 创建三个工人和一个守护线程, 模拟下班监督的工作场景。

```java
public static void main(String[] args) {
    Sweeper sweeper = new Sweeper();
    sweeper.start();
    Worker w1 = new Worker(new Date());
    sweeper.addWorker(w1);
    Worker w2 = new Worker(new Date());
    sweeper.addWorker(w2);
    Worker w3 = new Worker(new Date());
    sweeper.addWorker(w3);
    w1.start();
    w2.start();
```

```
        w3.start();
    }
```

代码测试结果如下，当工人都下班结束后，守护线程自动结束。

```
11 号员工 12:25:31 打卡上班
10 号员工 12:25:31 打卡上班
9 号员工 12:25:31 打卡上班
10 号员工不足 8 小时,不能下班...
11 号员工不足 8 小时,不能下班...
9 号员工不足 8 小时,不能下班...
10 号员工不足 8 小时,不能下班...
11 号员工不足 8 小时,不能下班...
9 号员工不足 8 小时,不能下班...
10 号员工不足 8 小时,不能下班...
11 号员工超过 8 小时,可以下班！
9 号员工不足 8 小时,不能下班...
11 号员工 20:25:31 下班------
10 号员工超过 8 小时,可以下班！
9 号员工超过 8 小时,可以下班！
9 号员工 20:25:31 下班------
10 号员工 23:25:31 下班------
```

1.7　本章习题

（1）以下哪个方法是 Runnable 接口中定义的方法？（　　）

A. start()　　　　B. run()　　　　C. stop()　　　　D. yield()

（2）以下哪个关于多线程的描述是正确的？（　　）

A. 多线程是 Java 语言独有的

B. 多线程运行必须要有多 CPU 支持

C. 多线程启动后，会马上得到 CPU 的运行

D. Java 语言支持多线程

（3）以下哪个方法可以启动一个新线程?（　　）

A. start()　　　　B. init()　　　　C. run()　　　　D. main()

（4）当一个处于阻塞状态的线程解除阻塞后，它将回到哪个状态？（　　）

A. 运行状态　　　B. 结束状态　　　C. 新建状态　　　D. 就绪状态

（5）线程的默认优先级是哪项？（　　）

A. 0　　　　　　B. 1　　　　　　C. 5　　　　　　D. 10

（6）以下代码的执行结果是（　　）。

```
class Example implements Runnable {
    public static void main(String args[]) {
        Example ex = new Example();
        Thread t = new Thread(ex);
        t.start();
    }
    void run() {
        System.out.print("pong");
    }
}
```

A. 输出 pong   B. 运行时输出异常信息
C. 运行后无任何输出   D. 编译失败

(7) 为了保证方法的线程安全，声明方法的时候需要使用哪个修饰符？（   ）

A. volatile   B. transient   C. synchronized   D. lock

(8) 如下代码的运行结果是（   ）。

```
public static void main(String[] args) {
    Thread t = new Thread(new Runnable() {
        public void run() {
            System.out.println("go");
        }
    });
    t.start();
    t.start();
}
```

A. go go   B. go 然后输出异常
C. go   D. 没有输出

(9) 如下代码的运行结果是（   ）。

```
public static void main(String[] args) {
    Thread t = new Thread(new Runnable() {
        public void run() {
            System.out.println("go");
        }
    });
    t.run();
}
```

A. 在主线程中输出 go   B. 在新线程中输出 go
C. 报错   D. 无输出

(10) 线程 A 访问正在被线程 B 用 synchronized 同步的共享资源对象时，会出现什么结

果？（　　）

  A. 进入 WAITING 等待状态　　　　B. 进入 BLOCKED 阻塞状态

  C. 无须等待，直接访问共享资源　　D. 不等待，退出

（11）调用 Object 对象的 wait()方法，会使当前线程进入什么状态？（　　）

  A. 阻塞等待状态　　　　　　　　　B. TIMED_WAITING 延时等待状态

  C. WAITING 等待状态　　　　　　　D. TERMINATED 结束状态

（12）通过调用线程的（　　）方法，可以使当前线程让出 CPU 的使用权。

  A. run　　　　　B. setPriority　　　C. yield　　　　D. sleep

（13）当（　　）方法结束时，线程进入终结状态。

  A. yield()　　　B. wait()　　　　　C. run()　　　　D. sleep()

（14）线程调用（　　）方法可以休眠一段时间，然后恢复运行。

  A. run　　　　　B. wait()　　　　　C. yield　　　　D. sleep

第 2 章 线程安全与共享资源竞争

本章主要介绍 Java 多线程中的同步机制，即当多个线程同时竞争访问共享资源时，如何保证线程安全。另外，还要考虑如何提高多线程并发访问的性能问题，这是一个非常棘手又非常重要的问题。

单线程就好比是问题域求解的单一对象，它每次只能做一件事情。因为只有一个实体对象，所以永远不用担心诸如 "两个对象试图同时使用同一个资源" 这样的问题。比如，两个人同时购买了同一个座位的火车票，且都购票成功；两个人同时使用同一个账户进行消费等。

Java 的多线程允许同时做多件事情。但是，两个及两个以上的线程彼此互相影响的问题也就出现了。如果不防范这种冲突，就可能发生两个用户购买了同一个座位的火车票，且都购票成功的事情。

2.1 synchronized 同步介绍

synchronized 要解决的是共享资源冲突的问题。假设一群人围在一起吃饭，你看好了盘里的一块肉，伸出筷子正要去夹的时候，那块肉却被人抢先夹走了。这多少让人有些懊恼，可你又无可奈何。怎样防止肉被别人抢走呢？

防止出现资源并发冲突的解决思路如下：当共享资源被任务使用时，要对资源提前加锁。所有任务都采用抢占模式，即某个任务会抢先对共享资源加上第一把锁。如果这是一个排他锁，其他任务在资源被解锁前就无法访问了。如果是共享锁，当浏览某些数据时，其他任务也可以同时浏览，但是不允许修改。

Java 提供了资源同步的关键字 synchronized，它的作用是获取指定对象的监视器锁。每个 Object 对象都内置了一个监视器锁，当某个线程获得了这个监视器锁后，其他线程再想获得这个对象的监视器锁，就必须要排队等待。也就是说：synchronized 关键字的底层，相当于一个排他锁。

2.2 synchronized 同步方法

多个线程同时访问同一个对象的某个方法时,如果该方法中存在对共享资源的操作,则可能引发线程安全问题。典型的共享资源有对象的成员变量、磁盘文件、静态变量等。

参见如下代码,这是一个倒计时器类,创建计时器对象时赋给初始值。调用 timeout() 方法就开始倒计时。

```java
public class Clock {
    private int beginer;
    public Clock(int beginer) {
        this.beginer = beginer;
    }
    public void timeout() {
        System.out.println(Thread.currentThread().getId() + ",进入...");
        while(this.beginer > 0) {
            beginer--;
            System.out.println(Thread.currentThread().getId()
                        + ",倒计时: " + beginer);
            try {
                Thread.sleep(1000);
            } catch (Exception e) {
            }
        }
    }
}
```

Clock 类中的 beginer 为共享资源。当计时器被一个线程调用时,倒计时效果显示正常:

```java
public static void main(String[] args) {
    Clock clock = new Clock(10);
    clock.timeout();
}
```

倒计时效果如下:

```
1,进入...
1,倒计时: 9
1,倒计时: 8
1,倒计时: 7
1,倒计时: 6
1,倒计时: 5
1,倒计时: 4
```

```
1,倒计时: 3
1,倒计时: 2
1,倒计时: 1
1,倒计时: 0
```

当两个线程同时使用一个计时器时，则会出现倒计时混乱现象，参见如下代码：

```
public static void main(String[] args) {
    Clock clock = new Clock(10);
    for(int i=0;i<2;i++) {
        new Thread(new Runnable() {
            public void run() {
                clock.timeout();
            }
        }).start();
    }
}
```

两个线程同时调用的效果如下（每次输出都会不同）：

```
8,进入...
8,倒计时: 9
9,进入...
9,倒计时: 8
9,倒计时: 7
8,倒计时: 6
9,倒计时: 4
8,倒计时: 4
8,倒计时: 2
9,倒计时: 2
9,倒计时: 0
8,倒计时: 0
```

为何两个线程使用同一个倒计时器，会频繁出现倒计时数字相同的现象呢？关键原因就是两个线程访问同一个计时器，计时器的成员变量 beginer 就成为多线程共享资源。在不使用锁策略时，两个线程经常会同时读取同一块内存地址，因此会频繁出现倒计时数相同的现象。

上述问题的解决方法如下：使用 synchronized 关键字，同步 timeout() 方法。即在同一时间，只允许一个线程进入 timeout() 方法。只有抢先进入 timeout() 方法的线程执行完毕，第二个线程才有机会进入 timeout() 方法。

方法 timeout() 的代码修改如下，增加 synchronized 同步策略：

```
public synchronized void timeout() {
```

```java
        System.out.println(Thread.currentThread().getId() + ",进入...");
        while(this.beginer > 0) {
            beginer--;
            System.out.println(Thread.currentThread().getId()+",倒计时:"+beginer);
            try {
                Thread.sleep(1000);
            } catch (Exception e) {
            }
        }
    }
}
```

增加 synchronized 同步方法后，代码测试结果如下。线程 8 先抢到了 Clock 对象的监视器锁，只有线程 8 倒计时完毕，退出 timeout()方法后，线程 9 才有机会获得监视器锁，并进入 timeout()方法。方法同步后，再也不会出现倒计时混乱现象了。

```
8进入...
8,倒计时: 9
8,倒计时: 8
8,倒计时: 7
8,倒计时: 6
8,倒计时: 5
8,倒计时: 4
8,倒计时: 3
8,倒计时: 2
8,倒计时: 1
8,倒计时: 0
9进入...
```

总结：上述计时器示例的关键是两个线程在同时使用同一个计时器，因此会出现共享资源竞争导致的计时错误现象。使用 synchronized 锁可以解决计时错误问题，但是只能有一个线程抢到计时器，另外一个线程排队进入 timeout()方法后，倒计时已经结束了。因此，想让两个线程都能正确地使用计时器，最好的办法当然是一个线程使用一个计时器，没有资源竞争，大家都会相安无事。

Clock 类不变，测试代码修改如下,即创建计时器的代码放到了循环内，每个线程使用自己的计时器，互不干扰。

```java
public static void main(String[] args) {
    for(int i=0;i<2;i++) {
        Clock clock = new Clock(10);
        new Thread(new Runnable() {
            public void run() {
                clock.timeout();
            }
```

```
        }).start();
    }
}
```

测试结果如下：

```
9,进入...
9,倒计时：9
8,进入...
8,倒计时：9
8,倒计时：8
9,倒计时：8
8,倒计时：7
9,倒计时：7
8,倒计时：6
9,倒计时：6
8,倒计时：5
9,倒计时：5
8,倒计时：4
9,倒计时：4
8,倒计时：3
9,倒计时：3
8,倒计时：2
9,倒计时：2
8,倒计时：1
9,倒计时：1
8,倒计时：0
9,倒计时：0
```

2.2.1　同步方法调用流程

synchronized 同步方法，并不是真的给方法加了锁，它的本质是使用了当前对象的监视器锁，其调用流程如下：

（1）线程 A 调用 clock 对象的 timeout()方法，进入方法体前，先试图获取当前 clock 对象的监视器锁。

（2）如果 clock 对象的监视器锁没有被占用，则线程 A 会获取 clock 对象的监视器锁，然后进入 timeout()方法；否则自旋等待获得锁（一般为抢占式，无须排队）。

（3）其他线程调用 timeout()方法时，执行顺序相同。

2.2.2　同步方法之间的互斥

提问：同一对象的同步方法之间是否会产生影响？

下面继续通过 Clock 类进行测试分析。

（1）在 Clock 类中新增一个正计时的方法 time()，这个 time()方法也是 synchronized 方法。为了与 timeout()方法对比，此处正计时只允许调用 10 次。

```java
public synchronized void time() {
    System.out.println(Thread.currentThread().getId() + ",进入time...");
    int end = this.beginer+1;
    while(end-beginer<10) {
        System.out.println(Thread.currentThread().getId() + ",正计时:" + end);
        try {
            Thread.sleep(1000);
        } catch (Exception e) {
        }
        end++;
    }
}
```

（2）time()与 timeout()方法都是 synchronized 方法，而且共享同一个成员变量 beginer。
（3）两个线程同时发出请求，分别调用同一个 clock 对象的 time()与 timeout()方法。

```java
public static void main(String[] args) {
    Clock clock = new Clock(10);
    new Thread(new Runnable() {
        public void run() {
            clock.time();
        }
    }).start();
    new Thread(new Runnable() {
        public void run() {
            clock.timeout();
        }
    }).start();
}
```

因为多线程任务是抢占式的，因此可能出现如下两种结果。执行结果 1：

```
8,进入time...
8,正计时：11
8,正计时：12
8,正计时：13
8,正计时：14
8,正计时：15
8,正计时：16
8,正计时：17
```

```
8,正计时: 18
8,正计时: 19
9,进入 timeout...
9,倒计时: 9
9,倒计时: 8
9,倒计时: 7
9,倒计时: 6
9,倒计时: 5
9,倒计时: 4
9,倒计时: 3
9,倒计时: 2
9,倒计时: 1
9,倒计时: 0
```

执行结果 2：

```
8,进入 timeout...
8,倒计时: 9
8,倒计时: 8
8,倒计时: 7
8,倒计时: 6
8,倒计时: 5
8,倒计时: 4
8,倒计时: 3
8,倒计时: 2
8,倒计时: 1
8,倒计时: 0
9,进入 time...
9,正计时: 1
9,正计时: 2
9,正计时: 3
9,正计时: 4
9,正计时: 5
9,正计时: 6
9,正计时: 7
9,正计时: 8
9,正计时: 9
```

总结：通过上述代码的反复测试可以发现，在某个时间 time() 与 timeout() 方法，只能有一个方法在运行，另外一个方法则处于等待状态。究其原因，就是因为执行 synchronized 方法前必须要获得对象的监视器，同一个对象的多个 synchronized 方法共享同一个对象监视器，因此我们可以简单总结为，同一个对象的 synchronized 方法之间是互斥的。即线程 A 调用 time() 方法，会阻碍线程 B 调用 timeout() 方法。

2.2.3 同步方法与非同步方法

提问：同一对象的同步方法与非同步方法之间是否会产生影响？

下面在 Clock 类中增加 tick()方法，即时钟的嘀嗒声。注意：这个方法为非同步方法。

```java
public void tick() {
    System.out.println(Thread.currentThread().getId() + ",进入tick...");
    while (true) {
        System.out.println(Thread.currentThread().getId() + ",tick");
        try {
            Thread.sleep(1000);
        } catch (Exception e) {
        }
    }
}
```

启动 3 个线程，分别调用 time()、timeout()、tick()方法：

```java
public static void main(String[] args) {
    Clock clock = new Clock(10);
    new Thread(new Runnable() {
        public void run() {
            clock.timeout();
        }
    }).start();
    new Thread(new Runnable() {
        public void run() {
            clock.time();
        }
    }).start();
    new Thread(new Runnable() {
        public void run() {
            clock.tick();
        }
    }).start();
}
```

程序运行结果如下：

```
8,进入timeout...
8,倒计时：9
10,进入tick...
10,tick
8,倒计时：8
```

```
10,tick
8,倒计时:7
10,tick
8,倒计时:6
10,tick
8,倒计时:5
10,tick
8,倒计时:4
10,tick
8,倒计时:3
10,tick
8,倒计时:2
10,tick
8,倒计时:1
10,tick
8,倒计时:0
10,tick
9,进入time...
9,正计时:1
10,tick
9,正计时:2
10,tick
9,正计时:3
10,tick
9,正计时:4
10,tick
9,正计时:5
10,tick
9,正计时:6
10,tick
9,正计时:7
10,tick
9,正计时:8
10,tick
9,正计时:9
10,tick
10,tick
```

分析程序的运行结果可以很容易得到结论,即同一个对象的 synchronized 方法之间是互斥的。例如,线程 8 和线程 9 准备执行同一个对象的 time() 和 timeout() 方法时,只能有一个方法被抢先执行,另一个则要等待。而 tick() 方法是非同步的方法,因此线程 10 可以一直运行,完全不需要受到锁的制约,自然也不会受到 synchronized 方法的影响。

Clock 案例在生活实践中也很好解释:一个计时器有倒计时和正计时两种功能,在同一时间只能使用其中一个功能。如果很多人都想使用这个计时器，则需要排队等待。而计时器的嘀嗒声，不受计时功能的影响。

2.3 synchronized 同步静态方法

前面我们讨论了使用 synchronized 同步成员方法带来的影响，下面我们探讨一下 synchronized 同步 static 方法。

2.3.1 单例高并发问题

单例(Singleton)是非常有名的设计模式，就是在程序运行期间，确保只能存在一个对象。其特点如下：

- 构造函数私有化（不允许类外 new 对象）。
- 单例对象使用 static 存储。
- 调用单例对象时使用静态方法。

```java
public class SingleObject {
    private static SingleObject obj;
    private SingleObject() {
    }
    public static SingleObject instance() {
        if(obj == null) {
            obj = new SingleObject();
            System.out.println(Thread.currentThread().getId()
                    + "，生成一个单例对象");
        }
        return obj;
    }
    public void doSomething() {
    }
}
```

如上单例代码，在高并发环境会出现冗余对象问题。例如，有 100 人第一次同时并发读取 Singleton，在没有同步控制的情况，多个线程可能同时进入 instance()方法，并判断 if (obj == null)成立，这样就会创建出多个单例对象。

可以使用如下代码模拟高并发环境，即使用线程池并发读取单例对象（线程池使用参见第 4 章讲解）。

```java
public static void main(String[] args) {
    ExecutorService pool = Executors.newCachedThreadPool();
    for(int i=0;i<100;i++) {
```

```
        pool.execute(new Runnable() {
            public void run() {
                new Thread(new Runnable() {
                    public void run() {
                        SingleObject obj = SingleObject.instance();
                        obj.doSomething();
                    }
                }).start();

            }
        });
    }
    pool.shutdown();
}
```

测试结果如下,虽然每次运行效果不同,但是可以轻易得到同时创建出多个单例的场景。

17,生成一个单例对象
19,生成一个单例对象
20,生成一个单例对象

为了解决单例的这个问题,可以在静态方法 instance 前使用 synchronized 进行同步。静态方法同步后,就不会再出现高并发环境下的冗余对象问题。

```
synchronized public static SingleObject instance() {}
```

2.3.2 类锁与对象锁

synchronized 同步成员方法,本质使用的是当前对象的监视器锁。而 synchronized 同步静态方法,则使用的是当前 Class 的监视器锁。

java.lang.Object 是所有对象的根,在 Object 的每个实例中都内置了对象监视器锁。
java.lang.Class 的父类也是 Object,在每个类型中也内置了一把锁,这与对象无关。

```
public final class Class<T> extends Object
        implements Serializable, GenericDeclaration, Type, AnnotatedElement
```

2.3.3 静态同步方法之间互斥

参见如下代码,在类 MyClass 中有两个静态方法 m1()和 m2(),都使用 **synchronized** 做了同步控制。

```
public class MyClass {
    synchronized public static void m1() {
```

```java
        System.out.println(Thread.currentThread().getId() + ",进入m1...");
        for(int i=0;i<5;i++) {
            System.out.println(Thread.currentThread().getId() + ",i=" + i);
            try {
                Thread.sleep(1000);
            } catch (Exception e) {
            }
        }
    }
    synchronized public static void m2() {
        System.out.println(Thread.currentThread().getId() + ",进入m2...");
        for(int i=0;i<5;i++) {
            System.out.println(Thread.currentThread().getId() + ",i=" + i);
            try {
                Thread.sleep(1000);
            } catch (Exception e) {
            }
        }
    }
}
```

创建两个线程，分别同时调用 MyClass 的 m1()和 m2()方法，测试代码如下：

```java
public static void main(String[] args) {
    new Thread(new Runnable() {
        public void run() {
            MyClass.m1();
        }
    }).start();
    new Thread(new Runnable() {
        public void run() {
            MyClass.m2();
        }
    }).start();
}
```

测试结果如下，可以看出 m1()与 m2()需要排队执行，即静态同步方法之间是互斥的。

```
8,进入m1...
8,i=0
8,i=1
8,i=2
8,i=3
8,i=4
```

```
9,进入m2...
9,i=0
9,i=1
9,i=2
9,i=3
9,i=4
```

2.3.4 静态同步方法与静态非同步方法

在 MyClass 中新增静态方法 m3()，这是一个没有使用 `synchronized` 同步的静态方法。

```java
public static void m3() {
    System.out.println(Thread.currentThread().getId() + ",进入m3...");
    for(int i=0;i<5;i++) {
        System.out.println(Thread.currentThread().getId() + ",i=" + i);
        try {
            Thread.sleep(1000);
        } catch (Exception e) {
        }
    }
}
```

创建三个线程，分别同时调用 MyClass 的 m1()、m2()、m3()方法，测试代码如下：

```java
public static void main(String[] args) {
    new Thread(new Runnable() {
        public void run() {
            MyClass.m1();
        }
    }).start();
    new Thread(new Runnable() {
        public void run() {
            MyClass.m2();
        }
    }).start();
    new Thread(new Runnable() {
        public void run() {
            MyClass.m3();
        }
    }).start();
}
```

测试结果如下，可以看出 m1()与 m2()需要排队执行，而 m3()的执行不受任何锁的影响。即同一个类的静态同步方法与静态非同步方法之间没有干扰。

```
10,进入 m3...
8,进入 m1...
10,i=0
8,i=0
10,i=1
8,i=1
10,i=2
8,i=2
10,i=3
8,i=3
10,i=4
8,i=4
9,进入 m2...
9,i=0
9,i=1
9,i=2
9,i=3
9,i=4
```

2.4　synchronized 同步代码块

synchronized 同步方法，就是线程在调用方法前获取对象监视器锁，方法执行完毕后就释放对象锁。

```
public class Clock {
    public synchronized void timeout() {...}
}
```

方法同步的关键是为了保护共享资源，如果 synchronized 方法中没有使用共享资源，就无须使用 synchronized 同步这个方法。

在同步方法中，使用共享资源的只是部分代码。为了提高并发性能，一般没必要在整个方法的运行期都持有监视器锁。

使用同步代码块模式，可以在方法中真正需要使用共享资源时再获取监视器锁，共享资源访问结束马上释放锁，这样就节省了对象监视器锁的占用时间，可以有效提高并发性。

2.4.1　锁当前对象

使用同步代码块模式，可以把 Clock 类的 timeout() 方法优化如下。优化之后的代码，对象监视器的使用时间更短，并发性能更高了。

```
public void timeout() {
```

```
        ...
        synchronized (this) {
             //调用共享资源
        }
        ...
}
```

synchronized (this) {}就是获取当前对象的监视器锁。**synchronized** 同步方法，本质就是隐式使用了 synchronized (this)。

2.4.2　锁其他对象

监视器锁内置于 Object 对象底层，所有对象的根都源于 Object，因此所有对象都有监视器锁。

使用其他 Object 对象的监视器锁，比使用自身对象的锁代码更加灵活。

修改 Clock 类，使用 Object 对象的监视器锁，同步效果相同。代码如下：

```
public class Clock {
    private Object lock;
    public Clock(int beginer) {
        lock = new Object();
    }
    public void time() {
        synchronized(lock) {
            ...
        }
    }
    public void timeout() {
        synchronized (lock) {
            ...
        }
    }
}
```

2.4.3　锁 Class

不仅每个对象内置了监视器锁，每个数据类型 Class 也内置了监视器锁。因此 Clock 类使用内置于 Class 中的锁，也可以很好地实现同步效果。

修改 Clock 类，使用类监视器锁，同步效果相同。代码如下：

```
public class Clock {
    public void time() {
        synchronized(Object.class) {
```

```
            ...
        }
    }
    public void timeout() {
        synchronized (Object.class) {
            ...
        }
    }
}
```

注意，**synchronized** (Object.class)使用的是类锁，不是对象锁。但是轻易不要使用 Object.**class** 的类锁，因为在整个项目中，如果其他业务模块也使用 Object.**class** 的类锁，这样就会产生并发冲突。合理使用类锁的基本原则：尽量使用当前类的监视器锁，如 Clock 类的同步模式可以进一步优化如下：

```
public class Clock {
    public void time() {
        synchronized(Clock.class) {
            ...
        }
    }
    public void timeout() {
        synchronized(Clock.class) {
            ...
        }
    }
}
```

2.5 项目案例：火车售票

在节假日，很多线路的火车票都是紧俏资源，为此也诞生了各种各样的抢票软件。火车票是允许多个窗口、多种渠道同时售卖的。因此行程、时间完全一致的某张火车票就成为了共享资源。

下面我们使用多线程模拟多个窗口同时出售火车票的场景（只考虑直达列车）。应使用同步机制，避免同一张票被成功出售多次。

2.5.1 共享任务模式

参见如下代码，TicketTask 代表售票任务，此处使用了同步代码块模式。注意：**synchronized** (**this**)这句代码，必须要位于 **while**(tickets>0)之后。如果移动到 **while**(tickets>0)之前，则所有的票只能被一个窗口售出。**if**(tickets>0) 这句代码也必须要有，否则会出现售票数量为负数的情况。

（1）定义售票任务，由于 ticket 是多线程共享资源，因此修改前需要使用 synchronized 同步块进行保护，否则并发售票的效果会出现错误值。

```java
public class TicketTask implements Runnable{
    private Integer ticket = 30;        //使用成员变量存储车票
    public void run() {
        while(this.ticket>0) {
            synchronized (this) {
                if(ticket>0){
                    System.out.println("窗口" + Thread.currentThread().getId()
                                    + "售出: " + ticket);
                    ticket--;
                }
            }
            try {
                Thread.sleep(100);
            } catch (Exception e) {
            }
        }
    }
}
```

（2）定义窗口类，每个窗口为一个线程。

```java
class Curtain extends Thread {
    private String cno;
    public String getCno() {
        return cno;
    }
    public Curtain(String cno,Runnable runnable) {
        super(runnable);        //给父类的任务赋值
        this.cno = cno;
    }
}
```

（3）使用多线程模拟多个窗口同时售票，这些窗口共享同一任务对象。注意：如果每个窗口都使用独立的任务对象，就不会出现共享资源竞争问题。TicketTask 的成员变量 ticket 是多线程共享的资源。

```java
public static void main(String[] args) {
    TicketTask task = new TicketTask();
    new Curtain("c01",task).start();
    new Curtain("c02",task).start();
```

```
        new Curtain("c03",task).start();
}
```

程序运行结果如下：

窗口 8 售出：30
窗口 9 售出：29
窗口 10 售出：28
窗口 8 售出：27
窗口 9 售出：26
窗口 10 售出：25
窗口 10 售出：24
窗口 9 售出：23
窗口 8 售出：22
窗口 9 售出：21
窗口 8 售出：20
窗口 10 售出：19
窗口 9 售出：18
窗口 10 售出：17
窗口 8 售出：16
窗口 9 售出：15
窗口 8 售出：14
窗口 10 售出：13
窗口 8 售出：12
窗口 9 售出：11
窗口 10 售出：10
窗口 9 售出：9
窗口 10 售出：8
窗口 8 售出：7
窗口 10 售出：6
窗口 9 售出：5
窗口 8 售出：4
窗口 10 售出：3
窗口 8 售出：2
窗口 9 售出：1

（4）要小心步骤（1）中的线程延迟代码 Thread.sleep(100)，这句代码放到 synchronized() 同步块中与放到同步块外，效果差别很大。

放在同步块中，当前线程执行完所有代码后，与排队线程再次产生竞争抢锁的关系，如果计算机速度快，可能出现所有任务被一个线程全部执行完成的效果。

延迟方法放在同步块外，在 while() 循环内，表明当前线程完成售票后会有短暂延迟，这样就不会和排队线程产生竞争关系，会出现多个线程交替执行的效果。

2.5.2 多任务模式

在 2.5.1 节的示例中,采用多个线程共享同一个任务的模式模拟多窗口售票。本节创建多个 TicketTask 任务,即每个线程使用自己的独立任务。

操作步骤如下:

(1) 定义 TicketTask 任务,此处的 ticket 不能使用成员变量,因为需要创建多个任务对象,这些任务需要共享票的数据,因此这里把 ticket 定义为静态变量。Object 也定义为静态变量,用于在多个任务之间共享同一把锁。

```java
public class TicketTask implements Runnable {
    static Integer ticket = 30;
    static Object obj = new Object();
    private String cno;
    public TicketTask(String cno) {
        this.cno = cno;
    }
}
```

(2) 在 TicketTask 类中定义成员方法,用于卖票。

```java
public void saleTicket() {
    if(ticket > 0) {
        System.out.println("窗口" + this.cno +","
            + Thread.currentThread().getId() + "售出 " + ticket);
        ticket --;
    }
}
```

(3) 在任务的 run() 方法中,锁定静态变量 obj,然后调用卖票方法。

```java
public void run() {
    while(ticket > 0) {
        //多个任务要使用同一把锁才能线程同步
        synchronized(obj) {
            saleTicket();
        }
        try {
            Thread.sleep(100);
        } catch (Exception e) {
        }
    }
}
```

（4）在主函数中创建三个线程，每个线程都使用独立的任务对象。

```java
public static void main(String[] args) {
    new Thread(new TicketTask("c1")).start();
    new Thread(new TicketTask("c2")).start();
    new Thread(new TicketTask("c3")).start();
}
```

程序运行结果如下：

窗口c3,10 售出 30
窗口c1,8 售出 29
窗口c2,9 售出 28
窗口c3,10 售出 27
窗口c1,8 售出 26
窗口c2,9 售出 25
窗口c3,10 售出 24
窗口c1,8 售出 23
窗口c2,9 售出 22
窗口c3,10 售出 21
窗口c2,9 售出 20
窗口c1,8 售出 19
窗口c1,8 售出 18
窗口c3,10 售出 17
窗口c2,9 售出 16
窗口c2,9 售出 15
窗口c3,10 售出 14
窗口c1,8 售出 13
窗口c3,10 售出 12
窗口c2,9 售出 11
窗口c1,8 售出 10
窗口c3,10 售出 9
窗口c1,8 售出 8
窗口c2,9 售出 7
窗口c2,9 售出 6
窗口c3,10 售出 5
窗口c1,8 售出 4
窗口c3,10 售出 3
窗口c2,9 售出 2
窗口c1,8 售出 1

2.5.3 共享车票资源

采用面向对象的编程模式，可以更加真实地模拟火车售票场景。

（1）定义火车票类 Ticket，每张车票有一个唯一的编号。

```java
public class Ticket {
    private String tno;
    public String getTno() {
        return tno;
    }
    public Ticket(String tno) {
        this.tno = tno;
    }
}
```

（2）定义售票任务，传入要销售的车票信息。

```java
public class TicketTask implements Runnable{
    private LinkedList<Ticket> tickets;
    public TicketTask(LinkedList<Ticket> tickets) {
        this.tickets = tickets;
    }
    public void run() {
        while(this.tickets.size() > 0) {
            synchronized(this) {
                Ticket tick = tickets.removeLast();
                System.out.println("窗口" + Thread.currentThread().getId()
                        + "售出: " + tick.getTno());
            }
            try {
                Thread.sleep(100);
            } catch (Exception e) {    }
        }
    }
}
```

（3）定义窗口类，每个窗口为一个线程。

```java
class Curtain extends Thread {
    private String cno;
    public String getCno() {
        return cno;
    }
    public Curtain(String cno,Runnable runnable) {
        super(runnable);     //给父类的任务赋值
        this.cno = cno;
    }
```

}

（4）使用多线程模拟多个窗口同时售票，多窗口共享任务对象，车票资源也共享。

```java
public static void main(String[] args) {
    LinkedList<Ticket> ticks = new LinkedList<>();
    for(int i=0;i<30;i++) {
        Ticket tick = new Ticket("t" + i);
        ticks.add(tick);
    }
    TicketTask task = new TicketTask(ticks);
    new Curtain("c1",task).start();
    new Curtain("c2",task).start();
    new Curtain("c3",task).start();
}
```

程序运行结果如下：

窗口 8 售出：t29
窗口 10 售出：t28
窗口 9 售出：t27
窗口 8 售出：t26
窗口 10 售出：t25
窗口 9 售出：t24
窗口 10 售出：t23
窗口 8 售出：t22
窗口 9 售出：t21
窗口 10 售出：t20
窗口 8 售出：t19
窗口 9 售出：t18
窗口 10 售出：t17
窗口 9 售出：t16
窗口 8 售出：t15
窗口 10 售出：t14
窗口 8 售出：t13
窗口 9 售出：t12
窗口 10 售出：t11
窗口 8 售出：t10
窗口 9 售出：t9
窗口 10 售出：t8
窗口 8 售出：t7
窗口 9 售出：t6
窗口 10 售出：t5
窗口 9 售出：t4

窗口 8 售出：t3
窗口 10 售出：t2
窗口 9 售出：t1
窗口 8 售出：t0

2.6　项目案例：家庭消费

家庭中有三位成员：父亲、妻子和儿子。这三个人共享同一个家庭账户。父亲负责赚钱，即向家庭账户中存钱；妻子和儿子负责消费。如果家庭账户中的余额不足以支持本次消费，那么应该提示余额不足。也就是说无论何时账户金额都不能为负。

家庭账户是典型的多线程共享资源，每个家庭成员则是一个独立的线程。为了确保在并发情况下账户的安全性，应该合理使用锁机制。

（1）定义家庭类 Family，Father（父亲）、Wife（妻子）、Son（儿子）为家庭成员。account 是家庭账户，name 为家庭名称。

```
public class Family {
    private String name;
    private Father father;
    private Wife wife;
    private Son son;
    private double account;
    public Family(String name){
        this.name = name;
    }
}
```

（2）定义 Father 类，它有一个唯一的行为 earning（挣钱）。

```
public class Father {
    private String name;
    private Family family;
    public Father(String name){
        this.name = name;
    }
    public void earning(double money){
        ...
    }
}
```

为了保证账户安全，查询账户和充值前，都需要使用 **synchronized** 锁定家庭对象。此处的账户 account 为基本类型，不能锁定。

```java
public void earning(double money){
    synchronized (this.family) {
        //先查询余额
        double leftMoney = this.family.getAccount();
        System.out.println("father 充值前，家庭余额： " + leftMoney);
        //充值
        this.family.setAccount(leftMoney + money);
        System.out.println("father 充值后，家庭余额："+ this.family.getAccount());
    }
}
```

（3）定义 Wife 类，它的唯一行为是 shopping（购物）。

```java
public class Wife {
    private String name;
    private Family family;
    public Wife(String name){
        this.name = name;
    }
    public void shopping(double money){
        ...
    }
}
```

购物行为需要并发操作账户，因此需要提前锁定 family 对象。当账户余额不足时，会暂停消费。

```java
public void shopping(double money){
    synchronized(this.family) {
        //先查询余额
        double leftMoney = this.family.getAccount();
        System.out.println("wife 消费前，家庭余额： " + leftMoney);
        //消费
        if(leftMoney >= money){
            this.family.setAccount(leftMoney-money);
            System.out.println("wife 消费后，家庭余额:"
                            + this.family.getAccount());
        }else{
            System.out.println("账户余额不足,wife 暂停消费*******");
        }
    }
}
```

（4）定义 Son 类，它的唯一行为是打游戏。

```java
public class Son {
    private String name;
    private Family family;
    public Son(String name){
        this.name = name;
    }
    public void playGame(double money){
        ...
    }
}
```

玩游戏的消费行为，同步锁定模式与 shopping 一致。

```java
public void playGame(double money){
    synchronized(this.family) {
        //先查询余额
        double leftMoney = this.family.getAccount();
        System.out.println("son 消费前，家庭余额: " + leftMoney);
        //消费
        if(leftMoney >= money){
            this.family.setAccount(leftMoney-money);
            System.out.println("son 消费后，家庭余额:" + this.family.getAccount());
        }else{
            System.out.println("账户余额不足，son 暂停消费*******");
        }
    }
}
```

（5）定义 FatherTask 任务，执行 Father 的 earning()行为。

```java
class FatherTask implements Runnable{
    private Father father;
    public FatherTask(Father father){
        this.father = father;
    }
    public void run() {
        while(true){
            this.father.earning(10);
            try {
                Thread.sleep(200);
            } catch (Exception e) {
            }
        }
    }
}
```

}

（6）定义 SonTask 任务,执行 Son 的 playGame()行为。

```java
class SonTask implements Runnable{
    private Son son;
    public SonTask(Son son){
        this.son = son;
    }
    public void run() {
        while(true){
            this.son.playGame(7);
            try {
                Thread.sleep(200);
            } catch (Exception e) {
            }
        }
    }
}
```

（7）定义 WifeTask 任务，执行 Wife 的 shopping 行为。

```java
class WifeTask implements Runnable{
    private Wife wife;
    public WifeTask(Wife wife){
        this.wife = wife;
    }
    public void run() {
        while(true){
            this.wife.shopping(8);
            try {
                Thread.sleep(200);
            } catch (Exception e) {
            }
        }
    }
}
```

（8）多线程模拟每一个家庭成员，运行程序，要保证账户在并发环境下的安全性。

```java
public static void main(String[] args) {
    Family family = new Family("三口之家");
    Father father = new Father("father");
    Wife wife = new Wife("wife");
    Son son = new Son("son");
```

```
        family.setFather(father);
        family.setWife(wife);
        family.setSon(son);
        new Thread(new FatherTask(father)).start();
        new Thread(new SonTask(son)).start();
        new Thread(new WifeTask(wife)).start();
}
```

运行效果如下，由于每次操作账户前都进行了线程同步锁定，因此账户数据始终正常。

```
father 充值前，家庭余额：0.0
father 充值后，家庭余额：10.0
wife 消费前，家庭余额：10.0
wife 消费后，家庭余额:2.0
son 消费前，家庭余额：2.0
账户余额不足，son 暂停消费*******
wife 消费前，家庭余额：2.0
账户余额不足，wife 暂停消费*******
father 充值前，家庭余额：2.0
father 充值后，家庭余额：12.0
son 消费前，家庭余额：12.0
son 消费后，家庭余额:5.0
son 消费前，家庭余额：5.0
账户余额不足，son 暂停消费*******
father 充值前，家庭余额：5.0
father 充值后，家庭余额：15.0
wife 消费前，家庭余额：15.0
wife 消费后，家庭余额:7.0
wife 消费前，家庭余额：7.0
账户余额不足，wife 暂停消费*******
father 充值前，家庭余额：7.0
 father 充值后，家庭余额：17.0
```

2.7　项目案例：别墅 Party

花园别墅举办一个 Party，必须要打开户外大门、穿过户外走廊，然后才能进入大厅。当然，如果不进入大厅，在别墅的花园中休息娱乐也可以。户外大门和大厅的门，为了安全，每次只允许一个人安检后进入。使用多线程，模拟多人顺序进入别墅，参加 Party 的场景。

2.7.1　无锁模式

新建别墅类 Villa，有两个方法，分别为进入户外大门 enterDoor()和进入大厅 enterHall()。

因为进入大门和进入大厅都需要耗费一些时间，因此方法中做了时间延迟。

```java
public class Villa {
    private String name;
    public Villa(String name) {
        this.name = name;
    }
    public void enterDoor(Person guest) {
        System.out.println(guest.getName() + "开始大门安检...");
        try {
            Thread.sleep(500);
        } catch (Exception e) {
        }
        System.out.println(guest.getName() + "已进入大门------");
    }
    public void enterHall(Person guest) {
        System.out.println(guest.getName() + "开始大厅安检....");
        try {
            Thread.sleep(500);
        } catch (Exception e) {
        }
        System.out.println(guest.getName()
                + "已进入大厅--------------------------");
    }
}
```

新建 Person 类，表示参加 Party 的人员。每个人员用一个线程进行模拟，因此需要实现 Runnable 接口。注意，进入别墅大门的人员，不一定都进入大厅，有些人可能留在花园里休息。因此这里使用随机数，只允许部分人员进入大厅。joinParty()方法，表示人员接到了参加 Party 的邀请，准备参加聚会。

```java
public class Person implements Runnable{
    private String name;
    private Villa villa;
    public String getName() {
        return name;
    }
    public Person(String name) {
        this.name = name;
    }
    public void joinPart(Villa villa) {
        this.villa = villa;
    }
```

```java
    public void run() {
        villa.enterDoor(this);                //进入大门
        int rand = (int)(Math.random()*100);
        if(rand>30) {
            try {
                Thread.sleep(500);
            } catch (Exception e) {
            }
                villa.enterHall(this);        //进入大厅
        }else {
            System.out.println(this.name + "在花园休息...");
        }
    }
}
```

测试代码如下，模拟多人同时参加 Party 的场景：

```java
public static void main(String[] args) {
    Villa villa = new Villa("鹰冠庄园");
    for(int i=0;i<8;i++) {
        Person p = new Person("p" + i);
        p.joinPart(villa);
        new Thread(p).start();
    }
}
```

测试结果如下，当前代码没有使用任何锁，因此会出现多个线程同时进入大门安检的现象，而我们项目的要求是进入大门和进入大厅必须是一个人一个人地顺序进入，不能并行，如 p0 开始大门安检与 p0 已进入大门之间不能有其他线程的输出信息。

```
p0 开始大门安检...
p1 开始大门安检...
p2 开始大门安检...
p3 开始大门安检...
p5 开始大门安检...
p4 开始大门安检...
p7 开始大门安检...
p6 开始大门安检...
p0 已进入大门------
p3 已进入大门------
p1 已进入大门------
p2 已进入大门------
p5 已进入大门------
p4 已进入大门------
```

p7 已进入大门------
p6 已进入大门------
p6 在花园休息...
p2 在花园休息...
p5 开始大厅安检...
p0 开始大厅安检...
p4 开始大厅安检...
p7 开始大厅安检...
p3 开始大厅安检...
p1 开始大厅安检...
p5 已进入大厅------------------------
p0 已进入大厅------------------------
p3 已进入大厅------------------------
p7 已进入大厅------------------------
p4 已进入大厅------------------------
p1 已进入大厅------------------------

2.7.2 单锁模式

因为别墅大门和大厅只允许一个人一个人地顺序进入，因此在并发编程中成为共享资源，需要使用同步机制进行控制。

修改 Villa 类，在 enterDoor()和 enterHall()方法前增加 **synchronized** 的同步控制。因为 **synchronized** 同步方法的本质是使用了当前对象的监视器锁，enterDoor()和 enterHall()方法使用的都是同一个监视器锁，因此称为单锁模式。

```java
public class Villa {
    synchronized public void enterDoor() {}
    synchronized public void enterHall() {}
}
```

测试代码不变，增加 synchronized 同步控制后，测试结果如下：

p0 开始大门安检...
p0 已进入大门------
p6 开始大门安检...
p6 已进入大门------
p5 开始大门安检...
p5 已进入大门------
p2 开始大门安检...
p2 已进入大门------
p1 开始大门安检...
p1 已进入大门------
p7 开始大门安检...

```
p7 已进入大门------
p4 开始大门安检...
p7 在花园休息....
p4 已进入大门------
p3 开始大门安检...
p3 已进入大门------
p4 开始大厅安检....
p3 在花园休息....
p4 已进入大厅------------------------
p1 开始大厅安检....
p1 已进入大厅------------------------
p2 开始大厅安检....
p2 已进入大厅------------------------
p5 开始大厅安检....
p5 已进入大厅------------------------
p6 开始大厅安检....
p6 已进入大厅------------------------
p0 开始大厅安检....
p0 已进入大厅------------------------
```

使用 synchronized 同步控制 enterDoor() 和 enterHall() 方法后，不会再出现多人同时进行大门安检或大厅安检的现象了。

但是，从反复测试的结果中是否发现了异常现象呢？所有准备进入大厅的人员，都排在了进入大门的行为之后，在某人"开始大门安检"和"已进入大门"之间是需要时间的，在这期间没有人能够进入大厅吗？

2.7.3 双锁模式

从 2.7.2 节的测试结果可以看到，所有准备进入大厅的人员，都排在了进入大门的行为之后。这貌似合理，其实存在问题。试想一下，客人进入大门后想要直奔大厅，但他却无法进入大厅，因为他要等所有人都进了大门后才能进入大厅，这是不是很滑稽？

出现上述现象的原因是单锁问题。enterDoor() 和 enterHall() 方法都使用了 synchronized 方法同步，两个方法默认使用了同一把锁（当前对象的内置监视器锁）。程序启动后，因为所有人的行为都是要先调用 enterDoor()，因此所有线程都排队在 enterDoor() 方法前等待拿到锁。即使是已经进入大门的线程，调用 enterHall() 申请的锁与调用 enterDoor() 申请的锁相同，自然排到了后面（由于锁是抢占式的，不能确保没有例外发生）。

进入大门与进入大厅，不应该存在互斥关系。为了使准备进入大门的线程与准备进入大厅的线程不产生影响，可以使用两把锁来解决问题。Villa 类修改如下：

```java
public class Villa {
    private Object lock1 = new Object();
```

```java
    private Object lock2 = new Object();
    public void enterDoor() {
        synchronized(lock1) {
            ...
        }
    }
    public void enterHall() {
        synchronized(lock2) {
            ...
        }
    }
}
```

测试代码不变，测试结果如下，在"p4 开始大门安检"与"p4 已进入大门"之间插入了"p0 开始大厅安检"的信息，说明 p0 进入大门后直奔大厅，p0 想进入大厅不会受其他想进入大门线程的任何干扰，它们是并行的业务行为，不是串行的。

p0 开始大门安检...
p0 已进入大门------
p5 开始大门安检...
p5 已进入大门------
p4 开始大门安检...
p0 开始大厅安检....
p4 已进入大门------
p6 开始大门安检...
p4 在花园休息....
p0 已进入大厅--------------------------
p5 开始大厅安检....
p6 已进入大门------
p6 在花园休息....
p7 开始大门安检...
p5 已进入大厅--------------------------
p7 已进入大门------
p3 开始大门安检...
p3 已进入大门------
p2 开始大门安检...
p7 开始大厅安检....
p7 已进入大厅--------------------------
p3 开始大厅安检....
p2 已进入大门------
p1 开始大门安检...
p1 已进入大门------
p3 已进入大厅--------------------------

```
p2 开始大厅安检....
p2 已进入大厅--------------------------
p1 开始大厅安检....
p1 已进入大厅--------------------------
```

2.8　JDK 常见类的线程安全性

在 JDK 中有很多 Class 并发与非并发的处理模式完全不同，有些类甚至是多次进行了修改，因此我们在使用时要非常小心。下面把常见的几个受多线程并发影响的类介绍一下。

2.8.1　集合 ArrayList 与 Vector

Vector 是早期应用最为普遍的一个动态集合，底层是静态数组结构。

```java
public class Vector<E> extends AbstractList<E>
        implements List<E>, RandomAccess, Cloneable, java.io.Serializable{
    protected Object[] elementData;
    protected int elementCount;
}
```

为了保证线程安全性，Vector 中的方法都使用 synchronized 同步模式。参见源代码如下：

```java
public class Vector {
    public synchronized void trimToSize() {}
    public synchronized void copyInto(Object[] anArray) {}
    public synchronized void ensureCapacity(int minCapacity) {}
    public synchronized void setSize(int newSize) {}
    public synchronized int capacity() {}
    public synchronized int size() {}
    public synchronized boolean isEmpty() {}
    public synchronized int indexOf(Object o, int index) {}
    public synchronized int lastIndexOf(Object o) {}
    public synchronized E elementAt(int index) {}
    public synchronized E lastElement() {}
    public synchronized void removeElementAt(int index) {}
    public synchronized void addElement(E obj) {}
    public synchronized Object clone() {}
    public synchronized Object[] toArray() {}
}
```

前面我们分析过，同一个对象的 synchronized 同步方法共享同一把对象锁，因此存在互斥现象。在并发环境，使用 Vector 集合，会严重影响并发性能。

JDK 为了优化 Vector 性能，新增了 ArrayList 类。ArrayList 的底层同 Vector 一致，都是静态数组结构。

```java
public class ArrayList<E> extends AbstractList<E>
            implements List<E>, RandomAccess, Cloneable,
                    java.io.Serializable
    transient Object[] elementData;
    private int size;
    private static final int DEFAULT_CAPACITY = 10;
}
```

参见 ArrayList 的源代码，为了提高性能，类中的方法都不再使用同步机制。

```java
public class ArrayList<E> extends AbstractList<E>
            implements List<E>, RandomAccess,
            Cloneable,java.io.Serializable{
    public void trimToSize() {}
    public void ensureCapacity(int minCapacity) {}
    public int size() {}
    public int indexOf(Object o) {}
    public E get(int index) {}
    public void add(int index, E element) {}
    public E remove(int index) {}
    public boolean remove(Object o) {}
}
```

由于 ArrayList 中没有使用 synchronized 同步机制，因此在并发访问集合时，可能会出现异常。参见如下代码测试：

```java
public static void main(String[] args) {
    ArrayList<Integer> arr = new ArrayList<>();
    new Thread(new Runnable() {
        public void run() {
            for(int i=0;i<10;i++) {
                arr.add(i);
                try {
                    Thread.sleep(200);
                } catch (Exception e) {
                }
            }
        }
    }).start();

    Iterator<Integer> it = arr.iterator();
```

```java
new Thread(new Runnable() {
    public void run() {
        while(it.hasNext()) {
            System.out.println(it.next());
        }
    }
}).start();
```

当一个线程向集合 ArrayList 中添加数据时，另外一个线程迭代访问集合中的数据，就会抛出 java.util.ConcurrentModificationException 异常。

Vector 与 ArrayList 在实际开发中，首先考虑性能。在没有并发安全问题时，优先推荐使用 ArrayList 集合。

2.8.2 StringBuffer 与 StringBuilder

当需要动态拼接字符串时，为了避免字符串常量碎片，通常使用 StringBuffer 或 StringBuilder 操作。

```java
public class StringBuffer extends AbstractStringBuilder
            implements java.io.Serializable, CharSequence {
    public synchronized int length() {}
    public synchronized int capacity() {}
    public synchronized void trimToSize() {}
    public synchronized char charAt(int index) {}
    public synchronized StringBuffer append(String str) {}
    public synchronized String substring(int start) {}
}
```

StringBuffer 中的方法，都使用了 synchronized 同步控制，因此在高并发环境会得到安全保证。而 StringBuilder 中的方法都没有使用同步控制。在没有并发安全问题时，优先推荐使用 StringBuilder 进行字符串操作。

```java
public class StringBuilder {
    public StringBuilder append(String str) {}
    public StringBuilder delete(int start, int end) {}
    public StringBuilder replace(int start, int end, String str) {}
    public StringBuilder insert(int offset, Object obj) {}
    public int indexOf(String str) {}
}
```

2.8.3 HashMap 与 ConcurrentHashMap

HashMap 的基本数据结构如图 2-1 所示，为数组+链表的结构。在 JDK1.8 中，当 key 值冲突数量超过 8，会把链表转换为红黑树。

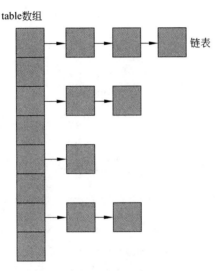

图 2-1 HashMap 数据结构

HashMap 中没有任何线程同步策略，因此在并发环境使用 HashMap 会抛出异常。为了解决 HashMap 的并发安全问题，增加了 ConcurrentHashMap 类。在 ConcurrentHashMap 类中大量使用了同步代码块，用于增加线程安全性。参见源代码如下：

```java
public class ConcurrentHashMap {
    final V putVal(K key, V value, boolean onlyIfAbsent) {
        synchronized(f) {
            ...
        }
    }
    final V replaceNode(Object key, V value, Object cv) {
        synchronized(f) {
            ...
        }
    }
    public void clear() {
        synchronized(f) {
            ...
        }
    }
}
```

上述代码中的 f 为图 2-1 中的节点，即 Node<K,V> f。为了线程安全且并发性能高，ConcurrentHashMap 锁定的是影响范围很小的节点对象，而不是整个数组或链表。

2.9 本章习题

（1）请问以下哪项关于锁的描述是正确的？（   ）
  A. 只有线程具有锁
  B. 所有类的对象实例都具有锁
  C. 基本数据类型也具有锁
  D. 只有 Runnable 对象具有锁
（2）对于如下代码，描述正确的是（   ）。

```java
public class Test {
    Object object = new Object();
    public void m1() {
        synchronized(object) {
            //同步代码块
        }
    }
}
```

  A. 当某个线程进入方法 m1() 的同步代码块后，在代码块运行结束前，其他线程无法进入这个同步代码块
  B. 某个线程调用 synchronized 锁定 object 对象后，在释放锁之前，其他线程无法访问 object 对象的方法
  C. object 对象可以同时使用多把锁
  D. object 对象内置了监视器锁，这是一个排他锁，当某个线程获得这个锁后，在释放前，其他线程无法获取 object 对象的这个锁
（3）对于如下代码，描述正确的是（   ）。

```java
public class Test {
    public void m1() {
        synchronized(this) {
            //同步代码块
        }
    }
    public void m2() {
        synchronized(this) {
            //同步代码块
        }
```

    }
}
```

A. 线程访问同步代码块前，需要获取当前对象的监视器锁
B. m1()方法的同步代码块与 m2()方法的同步代码块互不干扰
C. 同一个 Test 对象的 m1()方法与 m2()方法需要获取相同的监视器锁，因此具有互斥性
D. 不同对象的 m1()与 m2()方法，也具有干扰性

（4）对于如下代码，描述正确的是（　　）。

```
public class Test {
    public synchronized void m1() {
        //同步代码块
    }
    public void m2() {
        synchronized(this) {
            //同步代码块
        }
    }
}
```

A. m1()锁的是方法，m2()锁的是代码块。m1()的同步模式更加灵活、便捷
B. m1()的锁定模式，获取的也是当前对象的监视器锁，本质与 m2()的锁定模式一致
C. m2()的锁定模式，锁定共享资源的访问时间更短，比 m1()的同步模式灵活
D. m1()与 m2()方法，同步块之间具有互斥性

（5）对于如下代码，描述正确的是（　　）。

```
public class Test {
    public void m1() {
        synchronized(this) {
            //同步代码块
        }
    }
    public void m2() {
        synchronized(Test.class) {
            //同步代码块
        }
    }
    synchronized public static void m3() {
        //同步代码块
    }
}
```
```

A. 调用 m1() 方法需要获取当前 Test 对象的监视器锁，调用 m2() 方法需要获取 Test 类的监视器锁，这两种锁完全不同
B. m2() 方法与 m3() 方法使用的都是 Test 类内置的监视器锁
C. m1() 方法与 m2() 方法的锁具有互斥性
D. m3() 方法与 m2() 方法的锁具有互斥性

第 3 章 多线程通信

前面探讨了多线程的基础知识、多线程之间如何安全地访问共享资源等。这一章着重探讨一下多线程之间是如何交互的。

3.1 wait()与 notify()

java.lang.Object 类中内置了用于线程通信的方法 wait()、notify()与 notifyAll()。

```java
public class Object {
    public final void wait() throws InterruptedException {}
    public final native void wait(long timeout)
                throws InterruptedException;
    public final native void notify();
    public final native void notifyAll();
}
```

调用 Object 对象的 wait()方法，会导致当前线程进入 WAITING 等待状态，直到另外一个线程，调用该对象的 notify()或 notifyAll()方法，处于 WAITING 状态的线程才重新转为 RUNNABLE 状态。

调用对象的 wait()方法前，当前线程必须要拥有指定对象的监视器锁。调用 wait()方法后，会释放对象的监视器锁，然后当前线程进入 WAITING 状态。

```java
synchronized(obj) {
    obj.wait();
}
```

3.1.1 阻塞当前线程

参见如下代码，在主函数中启动一个线程，当 i=5 的时候，调用 object 对象的 wait()方法，这会导致当前线程 8 被阻塞（不是主线程被阻塞）。

```java
public static void main(String[] args) {
```

```
    Object object = new Object();
    new Thread(new Runnable() {
        public void run() {
            for(int i=0;i<10;i++) {
                System.out.println(Thread.currentThread().getId() + ",i=" +i);
                if(i==5) {
                    synchronized(object) {
                        try {
                            System.out.println(Thread.currentThread().getId()
                                             + "开始等待...");
                            object.wait();
                        } catch (Exception e) {
                            e.printStackTrace();
                        }
                    }
                }
            }
        }
    }).start();
}
```

程序运行结果如下，线程 8 被阻塞后，进入长期等待状态。

```
8,i=0
8,i=1
8,i=2
8,i=3
8,i=4
8,i=5
8 开始等待...
```

在主函数中，增加一个新的线程，延迟 1 秒后，发送 notify()通知。

```
new Thread(new Runnable() {
    public void run() {
        System.out.println(Thread.currentThread().getId() + " running...");
        try {
          Thread.sleep(1000);
            synchronized(object) {
                System.out.println(Thread.currentThread().getId()
                                 + ",发送 notify 通知...");
                object.notify();
            }
        } catch (Exception e) {}
```

        }
}).start();
```

重新运行程序，运行结果如下。线程 8 进入 WAITING 状态后，线程 9 延迟 1 秒后发出了 notify()通知。线程 8 接到通知后继续运行直至结束。

```
8,i=0
8,i=1
8,i=2
8,i=3
8,i=4
8,i=5
8 开始等待...
9 running...
9,发送 notify 通知...
8,i=6
8,i=7
8,i=8
8,i=9
```

注意，不管是调用 object 的 wait()方法，还是调用 object 的 notify()方法，都要提前获得 object 对象的监视器锁：

```
synchronized(object) {
    ...
}
```

3.1.2 案例分析：厨师与侍者 1

假设有一个小饭店，里面只有一个厨师和一个侍者（服务员）。厨师只有收到服务员的通知才开始做菜，没有工作时厨师就处于等待状态。服务员只有收到顾客的订单，才会通知厨师工作，没有订单时，服务员也处于等待状态。厨师做完菜，会通知服务员取餐。使用多线程，模拟这个小饭店的工作状况。

厨师什么时候做菜，应该做什么菜，由服务员根据订单通知厨师。而服务员什么时候可以取菜，则是看厨师什么时候做完菜，做完后才会通知服务员取菜。

操作步骤如下：

（1）新建订单类，模拟最多 10 个订单。

```java
class Order{
    private static int i=0;
    private int m_count;
    public Order(){
```
```

```java
            m_count = i++;
            if(m_count==10){
                System.out.println("没有食物了,结束!");
                System.exit(0);
            }
        }
    }
}
```

(2)新建饭店类,成员变量订单表示饭店当前是否存在订单。

```java
class Restaurant {
    public Order order;
}
```

(3)新建厨师类 Chef,Chef 启动后,就处于等待 waiter 通知的状态。饭店当前订单为空时,模拟生成新订单。

```java
class Chef extends Thread{
    private Restaurant restaurant;
    private Waiter waiter;
    public Chef(Restaurant restaurant,Waiter waiter){
        this.restaurant = restaurant;
        this.waiter = waiter;
    }
    public void run(){
        while(true){
            if(restaurant.order == null){
                restaurant.order = new Order();
                System.out.println("厨师-" + Thread.currentThread().getId()
                                    + ",接到新订单");
                synchronized(waiter){
                    System.out.println("厨师-" + Thread.currentThread().getId()
                                        + ",通知 waiter 取食物");
                    waiter.notify();
                }
                try {
                    Thread.sleep(1000);
                } catch (Exception e) {
                }
            }
        }
    }
} }
```

(4)新建侍者类 Waiter。厨师做完菜后,通知侍者取餐。

```java
class Waiter extends Thread{
    private Restaurant restaurant;
    public Waiter(Restaurant r){
        restaurant = r;
    }
    public void run() {
        while(restaurant.order == null){
            synchronized(this){
                try {
                    System.out.println("Waiter-" + Thread.currentThread().getId()
                            + ",等待中");
                    wait();
                    restaurant.order = null;
                    System.out.println("Waiter-" + Thread.currentThread().getId()
                            + ",收到通知,取走订单 ");
                } catch (Exception e) {    }
            }
} } }
```

（5）代码测试。

```java
public static void main(String[] args) {
    Restaurant restaurant = new Restaurant();
    Waiter waiter = new Waiter(restaurant);
    waiter.start();
    Chef chef = new Chef(restaurant,waiter);
    chef.start();
}
```

程序运行结果如下：

```
Waiter-8,等待中
厨师-9,接到新订单
厨师-9,通知 waiter 取食物
Waiter-8,收到通知,取走订单
Waiter-8,等待中
厨师-9,接到新订单
厨师-9,通知 waiter 取食物
Waiter-8,收到通知,取走订单
Waiter-8,等待中
厨师-9,接到新订单
厨师-9,通知 waiter 取食物
Waiter-8,收到通知,取走订单
Waiter-8,等待中
```

厨师-9,接到新订单
厨师-9,通知 waiter 取食物
Waiter-8,收到通知,取走订单
Waiter-8,等待中
厨师-9,接到新订单
厨师-9,通知 waiter 取食物
Waiter-8,收到通知,取走订单
Waiter-8,等待中
厨师-9,接到新订单
厨师-9,通知 waiter 取食物
Waiter-8,收到通知,取走订单
Waiter-8,等待中
厨师-9,接到新订单
厨师-9,通知 waiter 取食物
Waiter-8,收到通知,取走订单
Waiter-8,等待中
厨师-9,接到新订单
厨师-9,通知 waiter 取食物
Waiter-8,收到通知,取走订单
Waiter-8,等待中
厨师-9,接到新订单
厨师-9,通知 waiter 取食物
Waiter-8,收到通知,取走订单
Waiter-8,等待中
厨师-9,接到新订单
厨师-9,通知 waiter 取食物
Waiter-8,收到通知,取走订单
Waiter-8,等待中
没有食物了,结束!

3.1.3 案例分析:厨师与侍者 2

3.1.2 节的厨师与侍者案例引自 *Thinking in Java*,都是在厨师类中模拟创建的订单,然后通知服务员(侍者)取菜。这个流程过于简单了,与实际情况不符。本节在 3.1.2 节的基础上,做进一步优化。模拟顾客在饭店通过服务员点菜,然后服务员通知厨师做菜;厨师做菜完成后会通知服务员取菜;厨师与服务员无事时,则处于空闲等待状态。

实现步骤如下:
(1)定义订单类,设置订单编号和订单内容。

```java
public class Order implements Serializable{
    private String dno;
    private String info;
```

```java
    public String getDno() {
        return dno;
    }
    public String getInfo() {
        return info;
    }
    public void setInfo(String info) {
        this.info = info;
    }
    public Order(String dno) {
        this.dno = dno;
    }
}
```

（2）定义饭店类，模拟顾客进入饭店，随机生成订单。

```java
class Restaurant implements Runnable {
    private Waiter waiter;
    public void setWaiter(Waiter waiter) {
        this.waiter = waiter;
    }
    public void run() {
        while(true) {
            int rand = (int)(Math.random()*5000);
            try {
                System.out.println("饭店等待顾客中---------------");
                Thread.sleep(rand);
                String dno = "d" + System.currentTimeMillis();
                Order order = new Order(dno);
                order.setInfo("宫保鸡丁一份...");
                System.out.println("顾客来了，通知服务员点菜，生成订单：" + dno );
                synchronized (waiter) {
                    waiter.setOrder(order);   //把订单给服务员
                    waiter.setMsgID(1);
                    waiter.notify();
                }
            } catch (Exception e) {
            }
        }
    }
}
```

（3）创建 Waiter 类，它可以接收新订单通知，也可以接收取菜通知。

```java
class Waiter implements Runnable{
    private Order order;
    private Chef chef;
    private int msgID = 1; //如果是新订单通知 ID=1，如果是取菜通知 ID=2
    public void setMsgID(int msgID) {
        this.msgID = msgID;
    }
    public void setChef(Chef chef) {
        this.chef = chef;
    }
    public void setOrder(Order order) {
        this.order = order;
    }
    public void run() {
        while(true) {
            synchronized(this) {
                try {
                    System.out.println("服务员空闲等待中...");
                    this.wait();
                } catch (Exception e) {
                    e.printStackTrace();
                }
            }
            if(msgID == 1) {
                //服务员收到新订单通知
                System.out.println("waiter 收到订单: " + this.order.getDno()
                                    + "," + this.order.getInfo());
                //通知厨师做菜
                synchronized(chef) {
                    System.out.println("waiter 通知厨师做菜...");
                    chef.setOrder(order);
                    chef.notify();
                }
            }else {
                //服务员收到了取菜通知
                System.out.println("waiter 取菜给顾客....");
            }
        }
    }
}
```

（4）定义厨师类。厨师收到服务员的做菜通知后开始烹饪，菜做完后通知服务员取菜。

```java
class Chef implements Runnable{
    private Order order;
```

```java
    private Waiter waiter;
    public void setOrder(Order order) {
        this.order = order;
    }
    public void setWaiter(Waiter waiter) {
        this.waiter = waiter;
    }
    public void run() {
        while(true) {
            synchronized(this) {
                try {
                    System.out.println("厨师空闲等待中...");
                    this.wait();
                } catch (Exception e) {
                    e.printStackTrace();
                }
            }
            //厨师收到订单通知
            int rand = (int)(Math.random()*800);
            try {
                Thread.sleep(rand);
            } catch (Exception e) {
            }
            System.out.println("厨师做菜完成,通知waiter取菜...");
            synchronized(this.waiter) {
                waiter.setMsgID(2);
                waiter.notify();
            }
        }
    }
}
```

（5）在主函数中，模拟饭店的点菜流程，进行测试。

```java
public static void main(String[] args) {
    Chef chef = new Chef();
    Waiter waiter = new Waiter();
    waiter.setChef(chef);
    chef.setWaiter(waiter);
    Restaurant rest = new Restaurant();
    rest.setWaiter(waiter);
    new Thread(waiter).start();
    new Thread(chef).start();
    new Thread(rest).start();
}
```

程序运行结果如下,本节案例与真实项目环境更加贴近。

```
厨师空闲等待中...
服务员空闲等待中...
饭店等待顾客中---------------
顾客来了,通知服务员点菜,生成订单:d1597404292617
waiter 收到订单:d1597404292617,宫保鸡丁一份...
waiter 通知厨师做菜...
服务员空闲等待中...
饭店等待顾客中---------------
厨师做菜完成,通知 waiter 取菜...
厨师空闲等待中...
waiter 取菜给顾客...
服务员空闲等待中...
顾客来了,通知服务员点菜,生成订单:d1597404293698
饭店等待顾客中---------------
waiter 收到订单:d1597404293698,宫保鸡丁一份...
waiter 通知厨师做菜...
服务员空闲等待中...
厨师做菜完成,通知 waiter 取菜...
厨师空闲等待中...
```

3.1.4 案例分析:两个线程交替输出信息

本节案例是让两个线程交替输出信息:一个线程输出 aa,另一个线程输出 bb。

这个案例的关键在于共享对象的状态为开关项,当某个线程锁定对象并读取对象状态时,另一个线程必须等待。

操作步骤如下:

(1)新建状态类 State,设置布尔值 bRet 作为开关项。

```java
class State {
    public boolean bRet = false;
}
```

(2)新建任务类 PrintA,锁定状态对象,并判断开关项为真时进入等待状态。

```java
class PrintA implements Runnable {
private State state;
public PrintA(State state){
    this.state = state;
}
public void run(){
    while (true) {
```

```java
            try {
                synchronized (state) {
                    if(state.bRet){
                        state.wait();
                    }
                    System.out.println("aa...");
                    Thread.sleep(1000);
                    state.bRet = true;
                    state.notify();
                }
            }catch(Exception e){
                e.printStackTrace();
            }
}}}
```

（3）新建任务类 PrintB，锁定状态对象，并判断开关项为假时进入等待状态。

```java
class PrintB implements Runnable {
    private State state;
    public PrintB(State state){
        this.state = state;
    }
        public void run(){
            while (true) {
                try {
                    synchronized(state) {
                        if(!state.bRet){
                            state.wait();
                        }
                        System.out.println("bb...");
                        Thread.sleep(1000);
                        state.bRet = false;
                        state.notify();
                    }
                }catch(Exception e){
                    e.printStackTrace();
                }
        } } }
```

（4）代码测试。

```java
public static void main(String[] args) {
    State state = new State();
    new Thread(new PrintA(state)).start();
```

```java
        new Thread(new PrintB(state)).start();
}
```

测试结果如下，交替输出 aa...和 bb...，循环往复。

```
aa...
bb...
aa...
bb...
aa...
```

3.2 join 线程排队

线程 A 调用线程 B 对象的 join()方法，会导致线程 A 的运行中断，直到线程 B 运行完毕或超时，线程 A 才继续运行。

```java
public class Thread {
    public final void join()
            throws InterruptedException { }
    public final synchronized void join(long millis)
            throws InterruptedException {}
}
```

需要注意的是，join(long millis)方法中传入的超时参数不能为负数，否则将抛出 IllegalArgumentException 异常。如果 millis 参数为 0，则表示永远等待。

3.2.1 加入者与休眠者

分别定义加入者 Joiner 与休眠者 Sleeper。在 Sleeper 运行过程中，如果 Joiner 加入 Sleeper 的运行中，会导致 Sleeper 的运行被阻塞，直到 Joiner 运行完毕，Sleeper 才会继续运行。

开发步骤如下：

（1）新建类 Joiner，循环输出 10 个 k 值。

```java
class Joiner extends Thread{
    public void run() {
        System.out.println("Joiner 线程id="
                    + Thread.currentThread().getId()+ " run...");
        try {
            for(int i=0;i<10;i++){
                Thread.sleep(100);
                System.out.println("线程"
                        + Thread.currentThread().getId() + "---k=" + i);
```

```
            }
        } catch (Exception e) {
        }
        System.out.println(Thread.currentThread().getId()+ " end...");
    }
}
```

（2）新建类 Sleeper，循环输出 10 个 i 值。当 i=5 时，joiner 加入。

```
class Sleeper extends Thread{
    private Joiner joiner;
    public void setJoiner(Joiner joiner) {
        this.joiner = joiner;
    }
    public void run() {
        System.out.println("Sleeper 线程 id="
            + Thread.currentThread().getId() + " run...");
        try {
            for(int i=0;i<10;i++){
                if(i==5 && joiner != null) {
                    System.out.println("joiner 加入,线程" +
                                Thread.currentThread().getId() + "被阻塞");
                    joiner.join();
                }
                Thread.sleep(100);
                System.out.println("线程"
                            + Thread.currentThread().getId() + "---i=" + i);
            }
        } catch (Exception e) {   }
        System.out.println(Thread.currentThread().getId()+ " end...");
    }
}
```

（3）代码测试，同时启动 Joiner 与 Sleeper 两个线程。当 i=5 时，Joiner 加入 Sleeper，这将导致 Sleeper 的运行被阻塞。

```
public static void main(String[] args) {
    Sleeper sleeper = new Sleeper();
    Joiner joiner = new Joiner();
    sleeper.setJoiner(joiner);
    sleeper.start();
    joiner.start();
}
```

代码运行结果如下,在 Joiner 加入前,Joiner 与 Sleeper 两个线程同时运行。当 Joiner 加入后,Sleeper 被阻塞。直到 Joiner 运行结束,Sleeper 才重新被激活,并运行到结束。

```
Sleeper 线程 id=8 run...
Joiner 线程 id=9 run...
线程 8---i=0
线程 9---k=0
线程 8---i=1
线程 9---k=1
线程 8---i=2
线程 9---k=2
线程 8---i=3
线程 9---k=3
线程 8---i=4
joiner 加入,线程 8 被阻塞
线程 9---k=4
线程 9---k=5
线程 9---k=6
线程 9---k=7
线程 9---k=8
线程 9---k=9
9  end...
线程 8---i=5
线程 8---i=6
线程 8---i=7
线程 8---i=8
线程 8---i=9
8 end...
```

3.2.2 案例:紧急任务处理

在工作的过程中经常会遇到这种情况:当你正在处理日常工作时,突然之间来了紧急任务。这时你被迫停下手里的工作优先处理紧急任务,等到紧急任务处理完成后再继续刚才未完成的工作。

操作步骤如下:

(1)新建线程 Worker 代表日常任务的完成。当出现紧急任务时,日常任务被停止,直到紧急任务结束才继续。

```
class Worker extends Thread {
    private UrgentTask joiner;
    public void run() {
        int i = 0;
```

```java
        while(i < 9){
            try {
                if(i == 8){
                    UrgentTask urgent = joiner;
                    System.out.println("突然接到了紧急工作,需要去完成...");
                    urgent.start();
                    urgent.join();
                }
                System.out.println("我正在做日常工作,完成度"
                                    + (i * 10) + "%...");
                TimeUnit.SECONDS.sleep(1);
                i += 2;
            } catch (InterruptedException e) {
                e.printStackTrace();
            }
        }
        System.out.println("我已经完成了日常工作,完成度100%...");
    }
    public void setJoiner(UrgentTask joiner) {
        this.joiner = joiner;
    }
}
```

（2）新建类 UrgentTask，表示紧急任务。

```java
class UrgentTask extends Thread {
    public void run() {
        int i = 0;
        while(i < 9){
            try{
                System.out.println("紧急任务处理,完成度" + (i * 10) + "%+++");
                TimeUnit.SECONDS.sleep(1);
                i += 3;
            }catch(Exception e){
                e.printStackTrace();
            }
        }
        System.out.println("紧急工作完成度100%+++");
    }
}
```

（3）分别创建日常工作对象和紧急任务对象。当日常工作进行到一半时，紧急任务加入。

```java
public static void main(String[] args) {
    try {
        Worker worker = new Worker();
        UrgentTask j = new UrgentTask();
        worker.start();
        TimeUnit.SECONDS.sleep(3);
        worker.setJoiner(j);
    } catch (Exception e) {
        e.printStackTrace();
    }
}
```

程序运行结果如下：

```
我正在做日常工作,完成度 0%...
我正在做日常工作,完成度 20%...
我正在做日常工作,完成度 40%...
我正在做日常工作,完成度 60%...
突然接到了紧急工作,需要去完成...
紧急任务处理,完成度 0%+++
紧急任务处理,完成度 30%+++
紧急任务处理,完成度 60%+++
紧急工作完成度 100%+++
我正在做日常工作,完成度 80%...
我已经完成了日常工作,完成度 100%...
```

3.2.3 join 限时阻塞

join(long millis)就是指定阻塞时间,超时自动解锁。

修改 3.2.1 节中 Sleeper 的代码,原来的加入代码为 joiner.join(),现在修改为 joiner.join(300),即 Joiner 只阻塞 Sleeper 线程 300 ms。

其他代码不变,测试结果如下。从测试结果可以看出,Sleeper 被阻塞 300 ms 后继续运行,无须等到 Joiner 运行完毕。

```
Sleeper 线程 id=8 run...
Joiner 线程 id=9 run...
线程 8---i=0
线程 9---k=0
线程 8---i=1
线程 9---k=1
线程 8---i=2
线程 9---k=2
线程 8---i=3
```

```
线程 9---k=3
线程 8---i=4
joiner 加入,线程 8 被阻塞
线程 9---k=4
线程 9---k=5
线程 9---k=6
线程 9---k=7
线程 8---i=5
线程 9---k=8
线程 8---i=6
线程 9---k=9
9  end...
线程 8---i=7
线程 8---i=8
线程 8---i=9
8  end...
```

参考 join(long millis)方法的源码如下，join()方法的底层实现依赖的是 wait()方法。执行新线程 join()后，被阻塞的线程进入 WAITING 等待状态。join()延时结束后，通过 notify()通知，会唤醒处于 WAITING 状态的休眠线程。

```java
public final synchronized void join(long millis)
throws InterruptedException {
    long base = System.currentTimeMillis();
    long now = 0;
    if (millis < 0) {
        throw new IllegalArgumentException("超时值为负");
    }
    if (millis == 0) {
        while (isAlive()) {
            wait(0);
        }
    } else {
        while (isAlive()) {
            long delay = millis - now;
            if (delay <= 0) {
                break;
            }
            wait(delay);
            now = System.currentTimeMillis() - base;
        }
    }
}
```

3.3 线程中断

线程类 Thread 中的停止线程的 stop()方法与暂停线程的 suspend()方法已被废弃，如果想停止一个正在运行中的线程，可以尝试使用 interrupt()方法。

interrupt()方法并不能马上停止线程的运行，它只是给线程设置一个中断状态值，这相当于一个停止线程运行的建议，线程是否能够停止，由操作系统和 CPU 决定。

isInterrupted()方法用于判断当前线程是否处于中断状态。

```java
public class Thread implements Runnable {
    public final void stop() {}
    public final void suspend() {}
    public void interrupt() {}
    public boolean isInterrupted() {}
}
```

3.3.1 中断运行态线程

启动一个线程，延迟 500 ms 后，调用这个线程对象的 interrupt()方法。观察这个线程的中断状态，查看这个线程是否能够停止。

```java
public static void main(String[] args) {
    Thread t = new Thread(new Runnable() {
        public void run() {
            for(int i=0;i<10;i++) {
                if(Thread.currentThread().isInterrupted()) {
                    System.out.println("收到中断通知,结束线程...");
                    break;
                }else {
                    System.out.println(Thread.currentThread().getId()
                                        + ",i=" + i);
                    try {
                        Thread.sleep(100);
                    } catch (Exception e) {    }
                }
            }
        }
    });
    t.start();
    try {
      Thread.sleep(500);
    } catch (Exception e) {    }
```

```
        t.interrupt();
        System.out.println(t.getId() + "中断状态:" + t.isInterrupted());
}
```

反复运行上面的代码,可能得到完全不同的三种结果,分别如下:

(1)调用线程的 interrupt()方法后,最容易出现的结果就是中断状态为 false,也就是说 interrupt()方法没有起到任何效果。

```
8,i=0
8,i=1
8,i=2
8,i=3
8,i=4
8 中断状态:false
8,i=5
8,i=6
8,i=7
8,i=8
8,i=9
```

(2)调用线程的 interrupt()方法后,线程中断状态可能被设置为 true。但是线程仍然继续运行,并没有被停止。

```
8,i=0
8,i=1
8,i=2
8,i=3
8,i=4
8 中断状态:true
8,i=5
8,i=6
8,i=7
8,i=8
8,i=9
```

(3)调用线程的 interrupt()方法后,线程真的停止运行了。这种情况出现的次数较少,需要反复尝试才可以看到。

```
8,i=0
8,i=1
8,i=2
8,i=3
8,i=4
8 中断状态:true
```

收到中断通知，结束线程...

3.3.2 中断阻塞态线程

修改 3.3.1 节中的代码，让线程启动后进入 WAITING 状态。这时调用线程的 interrupt() 方法，观察程序运行状态。

```java
public static void main(String[] args) {
    Object obj = new Object();
    Thread t = new Thread(new Runnable() {
        public void run() {
            for(int i=0;i<10;i++) {
                if(Thread.currentThread().isInterrupted()) {
                    System.out.println("收到中断通知，结束线程...");
                    break;
                }else {
                    System.out.println(Thread.currentThread().getId()
                                    + ",i=" + i);
                    try {
                        Thread.sleep(100);
                        synchronized(obj) {
                            obj.wait();
                        }
                    } catch (Exception e) {
                        e.printStackTrace();
                    }
                }
            }
        }
    });
    t.start();
}
```

没有调用线程对象的 interrupt() 方法时，程序启动后进入 WAITING 状态。输出结果如下所示：

```
8,i=0
8 中断状态:false
```

调用线程对象的 interrupt() 方法，程序运行结果发生变化。

```java
public static void main(String[] args) {
    ...
```

```
        t.start();
        try {
            Thread.sleep(500);
        } catch (Exception e) {}
        t.interrupt();
        System.out.println(t.getId() + "中断状态:" + t.isInterrupted());
    }
```

程序运行结果如下，处于阻塞态的线程收到中断请求后会抛出 InterruptedException 异常。抛出异常后，中断状态被清除，程序继续向下运行。因此，打印异常信息后，会输出 8,i=1 的信息。

```
8,i=0
8 中断状态:false
java.lang.InterruptedException
    at java.lang.Object.wait(Native Method)
    at java.lang.Object.wait(Object.java:502)
    at com.icss.interrupt.InterruptTest2$1.run(InterruptTest2.java:18)
    at java.lang.Thread.run(Thread.java:745)
8,i=1
```

上面示例中断的是 WAITING 阻塞态的线程。如果线程是休眠阻塞态，也可中断线程运行。修改上面的代码如下，使线程进入休眠阻塞态。

```
    public static void main(String[] args) {
        Thread t = new Thread(new Runnable() {
            public void run() {
                for(int i=0;i<10;i++) {
                    if(Thread.currentThread().isInterrupted()) {
                        System.out.println("收到中断通知,结束线程...");
                        break;
                    }else {
                        System.out.println(Thread.currentThread().getId()
                                    + ",i=" + i);
                        try {
                            Thread.sleep(5000);
                        } catch (Exception e) {
                            e.printStackTrace();
                        }
                    }
                }
            }
        });
        t.start();
```

```
            try {
                Thread.sleep(500);
            } catch (Exception e) {
            }
            t.interrupt();
            System.out.println(t.getId() + "中断状态:" + t.isInterrupted());
}
```

程序运行结果如下:

```
8,i=0
8 中断状态:false
java.lang.InterruptedException: sleep interrupted
    at java.lang.Thread.sleep(Native Method)
    at com.icss.interrupt.InterruptTest3$1.run(InterruptTest3.java:14)
    at java.lang.Thread.run(Thread.java:745)
8,i=1
8,i=2
8,i=3
8,i=4
8,i=5
8,i=6
8,i=7
8,i=8
8,i=9
```

总结：处于 WAITING 阻塞态和休眠阻塞态的线程，调用 interrupt()方法会抛出 InterruptedException 异常。抛出异常后，中断状态被清除。但是，处于 BLOCKED 阻塞态的线程，调用 interrupt()方法不会产生中断影响。

3.3.3 如何停止线程

想停止一个正在运行的线程，最好的办法是设置一个布尔值 flag。在线程运行过程中，需要反复判断 flag 的值是否为真。当需要停止线程时，只要把 flag 的值设置为 false 即可。

```
public class StopTask implements Runnable{
    private boolean flag = true;
    public void run() {
        int i=0;
        while(flag){
            try {
                Thread.sleep(200);
            } catch (Exception e) {     }
            i++;
```

```
            System.out.println(Thread.currentThread().getId() +"--i=" + i);
        }
    }
    public void stop(){
        flag = false;
        System.out.println(Thread.currentThread().getId() + "发出停止命令...");
    }
}
```

调用 StopTask 的 stop()方法,可以马上停止线程的运行,参见如下代码:

```
public static void main(String[] args) {
    StopTask task = new StopTask();
    new Thread(task).start();
    try {
        Thread.sleep(1000);
        task.stop();
    } catch (Exception e) {    }
}
```

程序运行结果如下:

```
8--i=1
8--i=2
8--i=3
8--i=4
1 发出停止命令...
8--i=5
```

3.4　CountDownLatch 计数器

CountDownLatch 是 java.util.concurrent 包中的线程工具类,用于允许一个或多个线程阻塞等待,直到其他线程工作完毕才开始执行。

```
public class CountDownLatch {
    public CountDownLatch(int count) {}
    public void await() throws InterruptedException {}
    public void countDown() {}
    public long getCount() {}
}
```

CountDownLatch 类的构造函数,需要输入一个等待任务数量。每完成一项任务,就调用一次 countDown()方法。某个任务 A 启动后,调用 await()方法,进入阻塞等待状态。当调用 getCount()方法，发现等待完成任务数为 0 时，任务 A 开始执行最后的任务。

在现实生活中，任务等待现象很常见。例如，汽车厂商生产一辆汽车需要多种配件组装，包括车身、发动机、轮胎、底盘、驾驶室等。只有前面的工作都完成后，才能进入最后的总装环节。使用 CountDownLatch 可以模拟一个汽车的生产过程，最后总装出厂。

操作步骤如下：

（1）新建任务类 WorkingTask，每个 WorkingTask 完成后，都调用 CountDownLatch 的 countDown()方法表示任务已完成。

```java
public class WorkingTask implements Runnable {
    private CountDownLatch cl;
    public WorkingTask(CountDownLatch cl) {
        this.cl = cl;
    }
    public void run() {
        try {
            int r = (int)(Math.random()*10);
            Thread.sleep(r*1000);
            System.out.println(Thread.currentThread().getId() + "任务完成");
            cl.countDown();                    //任务完成后，计数器减一
        } catch (Exception e) {
            e.printStackTrace();
        }
    }
}
```

（2）新建汽车总装任务类 WaitingTask，它启动后就调用 CountDownLatch 的 await()方法进入阻塞等待状态。直到 CountDownLatch 的计数为 0，总装任务才继续执行。

```java
public class WaitingTask  implements Runnable{
    private CountDownLatch cl;
    public WaitingTask(CountDownLatch cl) {
        this.cl = cl;
    }
    public void run() {
        try {
            System.out.println(Thread.currentThread().getId()
                        + "阻塞等待其他任务完成...");
            cl.await();
            Thread.sleep(1000);
            System.out.println(Thread.currentThread().getId()
                        + "汽车总装完成，汽车出厂...");
        } catch (Exception e) {
```

```
            e.printStackTrace();
        }
    }
}
```

（3）在测试方法中，创建 CountDownLatch 对象，初始值为 10 个普通任务。

```
public static void main(String[] args) {
    CountDownLatch cl = new CountDownLatch(10);
    WaitingTask w = new WaitingTask(cl);
    new Thread(w).start();          //汽车总装任务
    for(int i=0;i<10;i++) {
        WorkingTask work = new WorkingTask(cl);
        new Thread(work).start();
    }
}
```

程序运行结果如下，总装任务最先启动，然后进入阻塞等待状态。其他任务都完成后，总装任务执行。

```
8 阻塞等待其他任务完成...
16 任务完成
9 任务完成
12 任务完成
17 任务完成
11 任务完成
18 任务完成
15 任务完成
13 任务完成
14 任务完成
10 任务完成
8 汽车总装完成，汽车出厂...
```

3.5　CyclicBarrier 屏障

CyclicBarrier 屏障适用于这种情况：你希望创建一组任务，它们并行地执行工作，然后在进行统一的下一步任务前彼此等待，直到所有的任务都完成。CyclicBarrier 屏障能让所有的并行任务都在 CyclicBarrier 屏障处停止，因此这组并行任务可以一致地向前移动。

CyclicBarrier 屏障与 CountDownLatch 计数器唯一的不同是：CountDownLatch 计数器只能触发一次，而 CyclicBarrier 屏障可以多次重用。

```
public class CyclicBarrier {
    public CyclicBarrier(int parties, Runnable barrierAction) {}
```

```java
    public int await()
            throws InterruptedException,
                BrokenBarrierException {}
    public int await(long timeout, TimeUnit unit)
            throws InterruptedException,
                BrokenBarrierException,
                TimeoutException {}
}
```

3.5.1 案例：矩阵分行处理

案例场景描述：每个工作线程处理矩阵的一行，然后等待屏障，直到所有行都被处理完才会执行跳闸任务（矩阵合并）。

（1）新建类 MatrixSolver，在构造函数中创建 CyclicBarrier 对象。

```java
public class MatrixSolver {
    final int num;        //线程参与者的数量
    final float[][] data;
    final CyclicBarrier barrier;
    public MatrixSolver(float[][] matrix) {
        data = matrix;
        num = matrix.length;
        //设置参与者数量和跳闸任务
        barrier = new CyclicBarrier(num, new Runnable() {
            public void run() {
                System.out.println("所有任务完成,"
                    + Thread.currentThread().getId() + "合并矩阵行...");
            }
        });
        for (int i = 0; i < num; i++) {
            Thread t = new Thread(new Worker(i));
            t.start();
        }
    } }
```

（2）新建内部类 Worker，每次任务执行完毕，都调用 CyclicBarrier 对象的 await()方法，在屏障前等待其他任务全部完成。

```java
public class MatrixSolver {
    ...
    class Worker implements Runnable {
        int myRow;
        Worker(int row) {
```

```java
            myRow = row;
        }
        public void run() {
            int rand = (int) (Math.random() * 1000);
            try {
                System.out.println("线程" + Thread.currentThread().getId()
                                    + "正在处理：第" + myRow + "行...");
                Thread.sleep(rand);
            } catch (Exception e) { }
            try {
                System.out.println("线程" + Thread.currentThread().getId()
                                    + "处理完成，栅栏前等待...");
                barrier.await();      //屏障前阻塞，等待跳闸
            } catch (Exception e) {}
        }
    }
}
```

（3）代码测试。

```java
public static void main(String[] args) {
    float[][] data = new float[10][5];
    MatrixSolver solver = new MatrixSolver(data);
}
```

程序运行结果如下：

```
线程 17 正在处理：第 9 行...
线程 16 正在处理：第 8 行...
线程 11 正在处理：第 3 行...
线程 12 正在处理：第 4 行...
线程 13 正在处理：第 5 行...
线程 15 正在处理：第 7 行...
线程 14 正在处理：第 6 行...
线程 8 正在处理：第 0 行...
线程 9 正在处理：第 1 行...
线程 10 正在处理：第 2 行...
线程 10 处理完成，栅栏前等待...
线程 11 处理完成，栅栏前等待...
线程 8 处理完成，栅栏前等待...
线程 16 处理完成，栅栏前等待...
线程 17 处理完成，栅栏前等待...
线程 12 处理完成，栅栏前等待...
线程 14 处理完成，栅栏前等待...
线程 15 处理完成，栅栏前等待...
```

线程 13 处理完成，栅栏前等待...
线程 9 处理完成，栅栏前等待...
所有任务完成，9 合并矩阵行...

3.5.2 案例：赛马游戏

案例场景描述：赛马一类的游戏，为了保证赛制公平，规定只有当所有的参赛马都到了指定的栏杆处才能开始进行比赛。

（1）新建类 HorseGame，在构造函数中创建 CyclicBarrier 对象。

```java
public class HorseGame {
    CyclicBarrier barrier;
    public HorseGame(int num) {
        barrier = new CyclicBarrier(num, new Runnable() {
            public void run() {
                System.out.println("所有参赛马匹准备就绪，开始比赛...");
            }
        });
        for (int i = 0; i < num; i++) {
          Thread t = new Thread(new HorseTask(i));
          t.start();
        }
    }
}
```

（2）新建内部类 HorseTask 表示赛马的准备工作。准备工作完成后，调用 CyclicBarrier 对象的 await()方法，等待其他赛马就绪。

```java
class HorseTask implements Runnable{
    int hno;
    public HorseTask(int hno) {
        this.hno = hno;          //马匹编号
    }
    public void run() {
        System.out.println(Thread.currentThread().getId()
                        + ": 赛马" + hno + "进场...");
        int rand = (int) (Math.random() * 1000);
        try {
            Thread.sleep(rand);
        } catch (Exception e) {
        }
        System.out.println(Thread.currentThread().getId() +
                        ": 赛马" + hno + "准备就绪...");
        try {
```

```
            barrier.await();        //栏杆前阻塞，等待其他赛马就绪
        } catch (Exception e) {
        }
    }
}
```

（3）代码测试。

```
public static void main(String[] args) {
    HorseGame race = new HorseGame(8);
}
```

运行结果如下：

```
8：赛马 0 进场...
9：赛马 1 进场...
12：赛马 4 进场...
11：赛马 3 进场...
13：赛马 5 进场...
15：赛马 7 进场...
10：赛马 2 进场...
14：赛马 6 进场...
12：赛马 4 准备就绪...
13：赛马 5 准备就绪...
15：赛马 7 准备就绪...
8：赛马 0 准备就绪...
9：赛马 1 准备就绪...
11：赛马 3 准备就绪...
10：赛马 2 准备就绪...
14：赛马 6 准备就绪...
所有参赛马匹准备就绪，开始比赛...
```

3.6 Exchanger

Exchanger 类是在两个任务之间交换对象的双向通道。当任务进入通道时，它们各自拥有一个对象，当它们离开时，它们都拥有之前由对方持有的对象。

Exchanger 的典型应用场景：一个任务在创建对象，而另一个任务在消费这些对象。通过 Exchanger 进行对象交换，可以使对象刚创建完成就马上被消费。

当线程调用 Exchanger.exchanger()方法时，它将阻塞直到另一个线程调用它自身的 exchanger()方法，那时，这两个 exchanger()方法将全部完成，而两个线程之间也随之交换了对象。

案例场景描述：在 NBA 球队中，存在着球队之间的球员互换。球员互换，必须是两个

球队之间达成协议，转让一个球员出去，得到另外一个球员进来。

（1）两个线程调用相同的 Exchanger 对象的 exchange()方法，进行数据互换。如果只有一个线程调用 exchange()方法，找不到互换线程，则任务阻塞等待。

```java
static void trade(String team, String data, Exchanger exchanger) {
    try {
        System.out.println(Thread.currentThread().getId()
                            + "," + team + "准备转让" + data);
        int rand = (int)(Math.random() * 10000);
        Thread.sleep(rand);
        String info = (String) exchanger.exchange(data);
        System.out.println(Thread.currentThread().getId()
                            + "," + team + "得到了" + info);
    } catch (InterruptedException e) {
        e.printStackTrace();
    }
}
```

（2）使用 4 个线程，模拟 4 支球队的球员互换，线程之间的配对自动完成。

```java
public static void main(String[] args) {
    final Exchanger exchanger = new Exchanger();
    new Thread() {
        public void run() {
            trade("火箭队", "姚明", exchanger);
        }
    }.start();
    new Thread() {
        public void run() {
            trade("马刺队","托尼-帕克", exchanger);
        }
    }.start();
    new Thread() {
        public void run() {
            trade("小牛队","韦斯利-马修斯", exchanger);
        }
    }.start();
    new Thread() {
        public void run() {
            trade("勇士队", "斯蒂芬-库里", exchanger);
        }
    }.start();
}
```

程序运行结果如下（反复运行，配对结果不同）：

8,火箭队准备转让姚明
9,马刺队准备转让托尼-帕克
10,小牛队准备转让韦斯利-马修斯
11,勇士队准备转让斯蒂芬-库里
11,勇士队得到了姚明
8,火箭队得到了斯蒂芬-库里
9,马刺队得到了韦斯利-马修斯
10,小牛队得到了托尼-帕克

(3) 使用 3 个线程，模拟 3 支球队的球员互换，则只能有一对互换成功，另一支球队阻塞等待。

```java
public static void main(String[] args) {
    final Exchanger exchanger = new Exchanger();
    new Thread() {
        public void run() {
            trade("火箭队", "姚明", exchanger);
        }
    }.start();
    new Thread() {
        public void run() {
            trade("马刺队","托尼-帕克", exchanger);
        }
    }.start();
    new Thread() {
        public void run() {
            trade("小牛队","韦斯利-马修斯", exchanger);
        }
    }.start();
}
```

运行结果如下，线程 9 无法配对，会长期阻塞等待：

8,火箭队准备转让姚明
9,马刺队准备转让托尼-帕克
10,小牛队准备转让韦斯利-马修斯
8,火箭队得到了韦斯利-马修斯
10,小牛队得到了姚明

3.7　Semaphore 信号灯

Semaphore 信号灯持有一组许可证，在 Semaphore 的构造函数中，输入许可证的数量。某个线程工作前，调用 Semaphore 对象的 acquire()获取许可证，然后工作。如果许可证已被其他线程占用，则阻塞等待其他线程释放许可证后再工作。线程工作完成，需要调用 release()方法释放许可证。

```java
public class Semaphore
            implements java.io.Serializable {
    public Semaphore(int permits) {}
    public void acquire()
            throws InterruptedException {}
    public void release() {}
}
```

案例场景描述：现在有多名学生要做实验，但是实验设备只有 3 台。学生必须要排队等待，按照顺序做完实验。

（1）新建学生任务类 StudentTask，做实验前需要调用 Semaphore 的 acquire()方法获取许可证。实验完毕，必须要释放许可证。

```java
class StudentTask implements Runnable{
    private Semaphore smp;
    private String sno;
    public StudentTask(Semaphore smp,String sno) {
        this.smp = smp;
        this.sno = sno;
    }
    public void run() {
        try {
            smp.acquire();   //排队等待获取许可证
            //得到许可证后，开始做实验
            int rand = (int)(Math.random()*5000);
            Thread.sleep(rand);
            System.out.println(Thread.currentThread().getId() + ","
                            + sno + "实验完毕" );
        } catch (Exception e) {
        }finally {
            smp.release();       //释放许可证
        }
    }
}
```

（2）3 个许可证，表示 3 台实验机。模拟 10 名学生排队做实验。

```java
public static void main(String[] args) {
    Semaphore smp = new Semaphore(3);
    for(int i=0;i<10;i++) {
        StudentTask s = new StudentTask(smp,"s"+i);
        new Thread(s).start();
    }
}
```

程序运行结果如下，10 名学生顺序完成了实验：

8,s0 实验完毕
10,s2 实验完毕
9,s1 实验完毕
11,s3 实验完毕
12,s4 实验完毕
13,s5 实验完毕
14,s6 实验完毕
16,s8 实验完毕
15,s7 实验完毕
17,s9 实验完毕

3.8 死锁

某个线程在等待另一个线程结束，而后者又在等待其他线程。如果出现线程间的循环等待，就可能出现死锁现象。死锁是个棘手的问题，JDK 在 1.5 版以前多次做过线程类的底层方法变更，主要原因就是防止死锁，提高并发效率。

死锁的发生需要同时满足以下四个必要条件。

（1）互斥条件：多线程使用的资源中至少有一个不能共享，如一根筷子一次只能被一人使用。

（2）线程持有资源并等待另一个资源：如拿着一根筷子的同时等待另一根筷子。

（3）资源不能被抢占：如不能抢别人的筷子。

（4）循环等待资源：一直不退出。

因为要发生死锁的前提是同时满足以上四个条件，所以要防止死锁的话，只需要破坏其中一个即可。Java 对死锁并没有提供语言层面上的底层支持，能否通过仔细的程序设计来避免死锁，这取决于程序员自身。

有死锁隐患的程序在相同的环境下运行，不是必然发生死锁现象，这给程序的测试、调试都带来很大的困扰。

3.8.1 案例：银行转账引发死锁

银行账户之间可以进行转账，下面我们演示由于设计不周全，导致银行转账死锁的案例。操作步骤如下。

（1）新建银行账户实体类，它有账号和余额两个属性。

```java
public class Account {
    private String id;       //账号
    private double money;    //余额
    public Account(String id) {
        this.id = id;
    }
    public String getId() {
        return id;
    }
    public double getMoney() {
        return money;
    }
    public void setMoney(double money) {
        this.money += money;
    }
}
```

（2）新建银行类 Bank，业务方法为转账 transferMoney()，为了转账安全，从一个账户转到另外一个账户时，需要使用 synchronized 锁定账户对象。

```java
public class Bank {
    public void transferMoney(Account from, Account to, double money) {
        try {
            synchronized(from) {
                if (from.getMoney() - money < 0) {
                    System.out.println("金额不足，转账失败");
                }
                from.setMoney(-money);
                synchronized(to) {
                    to.setMoney(money);
                }
                System.out.println(from.getId() + "转账成功,"
                        + to.getId() + "收到金额" + money + "元");
            }
        } catch (Exception e) {
            e.printStackTrace();
        }
    }
```

        }
    }

（3）代码测试，两个账户同时进行转账，a2 转账到 a3，而 a3 转账到 a2。

```
public static void main(String[] args) {
    final Bank bank = new Bank();
    final Account a2 = new Account("6222600260001072101");
    a2.setMoney(1000);
    final Account a3 = new Account("6222600260001085202");
    a3.setMoney(1000);
    new Thread(){
        public void run(){
            bank.transferMoney(a2, a3, 50);
        }
    }.start();
    new Thread(){
        public void run(){
            bank.transferMoney(a3, a2, 100);
        }
    }.start();
}
```

程序运行结果如下，并未出现死锁：

```
6222600260001072101 转账成功,6222600260001085202 收到金额 50.0 元
6222600260001085202 转账成功,6222600260001072101 收到金额 100.0 元
```

（4）在 Bank 的 transferMoney() 中增加一句转账延迟。

```
public void transferMoney(Account from, Account to, double money) {
    try {
        synchronized(from) {
            Thread.sleep(1000);     //转账延迟
            if (from.getMoney() - money < 0) {
                System.out.println("金额不足，转账失败");
            }
            from.setMoney(-money);
            synchronized(to) {
                to.setMoney(money);
            }
            System.out.println(from.getId() + "转账成功,"
                    + to.getId() + "收到金额" + money + "元");
        }
    } catch (Exception e) {
```

```
            e.printStackTrace();
        }
    }
```

（5）再次进行转账测试，就会出现死锁现象，即程序始终处于运行状态，但是却没有任何内容输出。

仅仅是加了一句 Thread.sleep(1000)语句，就可以让两个账户向对方转账的行为同时发生，从而引发了死锁。

解决这样的问题可能需要去制定锁的顺序（唯一标识、哈希值等），并且在整个应用程序中按照这个顺序去执行。在这个案例中，针对银行账户都会有一个唯一且固定的卡号，我们通过对银行卡号的字典顺序进行比较，进而确定锁的顺序。

（6）通过账号的字典顺序，确定转账顺序，转账代码调整如下。

```java
public void transferMoney(Account from, Account to, double money) {
    try {
        int result = from.getId().compareTo(to.getId());
        if(result>0) {
            synchronized(from) {
                Thread.sleep(1000);      //转账延迟
                from.setMoney(-money);
                synchronized(to) {
                    to.setMoney(money);
                }
                System.out.println(from.getId() + "转账成功,"
                        + to.getId() + "收到金额" + money + "元");
            }
        }else {
            synchronized(to) {
                Thread.sleep(1000);      //转账延迟
                to.setMoney(money);
                synchronized(from) {
                    from.setMoney(-money);
                }
                System.out.println(from.getId() + "转账成功,"
                        + to.getId() + "收到金额" + money + "元");
            }
        }
    } catch (Exception e) {
        e.printStackTrace();
    }
}
```

调整转账顺序后，代码逻辑没有发生变化，但是不会再发生死锁现象。程序运行结果如下：

6222600260001085202 转账成功,6222600260001072101 收到金额 100.0 元
6222600260001072101 转账成功,6222600260001085202 收到金额 50.0 元

3.8.2 案例：哲学家就餐死锁

案例场景描述：哲学家就餐问题是一个经典的死锁案例。一群哲学家围坐一桌就餐，与普通的就餐场景不同的是，两个哲学家之间只有一根筷子。哲学家要做的事情就是吃菜、思考。哲学家每次夹菜时，都遵循先拿左手的筷子，再拿右手的筷子，左右手筷子都拿到后，就可以夹菜了。如果左手或右手的筷子被别人占用，则哲学家需要等待。在等待右手筷子时，左手的筷子并不释放。

代码实现步骤如下：

（1）新建筷子类 ChopStick，每做一次 new ChopStick()，则筷子的编号加 1。

```java
class ChopStick{
    private static int counter = 0;
    private int number = counter++;
    public String toString(){
        return "筷子: " + number;
    }
}
```

（2）新建哲学家类 Philosopher，创建哲学家对象时需要设置其左右手的筷子对象。哲学家的行为就是思考、吃菜，循环往复。

```java
class Philosopher extends Thread{
    private static int counter = 0;
    private int number = counter++;
    private ChopStick leftChopstick;
    private ChopStick rightChopstick;
    public Philosopher(ChopStick left,ChopStick right){
        leftChopstick = left;
        rightChopstick = right;
        start();
    }
    public void think(){
        try {
            sleep(33);                    //思考时间越短，越容易发生死锁
        } catch (Exception e) {
        }
```

```java
    }
    public void eat(){
        synchronized(leftChopstick){
            System.out.println(Thread.currentThread().getId() + "持有"
                        + leftChopstick + " 等待 " + rightChopstick);
            synchronized(rightChopstick){
                System.out.println(this + " ---吃菜... ");
            }
        }
    }
    public String toString(){
        return "哲学家: " + number;
    }
    public void run(){
        while(true){
            think();
            eat();
        }
    }
}
```

（3）代码测试，创建 5 个哲学家，调整思考时间（思考时间越短，越容易发生死锁），一定会出现死锁现象。

```java
public static void main(String[] args) {
    Philosopher[] philosophers = new Philosopher[5];//人越少，越容易发生死锁
    ChopStick left,right,first;
    left = new ChopStick();
    right = new ChopStick();
    first = left;
    for(int i=0;i<philosophers.length-1;i++){
        philosophers[i] = new Philosopher(left,right);
        left = right;            //吃饭时先取左边的筷子，再取右边的筷子
        right = new ChopStick();
    }
    //最后一人的右手筷子为 first（转了一圈）
    philosophers[philosophers.length-1] = new Philosopher(left,first);
}
```

程序运行结果如下，每次死锁的时间不确定，但一定会发生死锁：

哲学家: 1 ---吃菜...
11 持有筷子: 3 等待 筷子: 4
哲学家: 3 ---吃菜...

12 持有筷子：4 等待　筷子：0
哲学家：4 ---吃菜...
8 持有筷子：0 等待　筷子：1
哲学家：0 ---吃菜...
10 持有筷子：2 等待　筷子：3
哲学家：2 ---吃菜...
9 持有筷子：1 等待　筷子：2
哲学家：1 ---吃菜...
12 持有筷子：4 等待　筷子：0
11 持有筷子：3 等待　筷子：4
哲学家：4 ---吃菜...
哲学家：3 ---吃菜...
8 持有筷子：0 等待　筷子：1
哲学家：0 ---吃菜...
10 持有筷子：2 等待　筷子：3
哲学家：2 ---吃菜...
9 持有筷子：1 等待　筷子：2
哲学家：1 ---吃菜...
12 持有筷子：4 等待　筷子：0
11 持有筷子：3 等待　筷子：4
哲学家：4 ---吃菜...
哲学家：3 ---吃菜...
8 持有筷子：0 等待　筷子：1
哲学家：0 ---吃菜...
10 持有筷子：2 等待　筷子：3
哲学家：2 ---吃菜...
9 持有筷子：1 等待　筷子：2
哲学家：1 ---吃菜...
12 持有筷子：4 等待　筷子：0
哲学家：4 ---吃菜...
11 持有筷子：3 等待　筷子：4
哲学家：3 ---吃菜...
8 持有筷子：0 等待　筷子：1
哲学家：0 ---吃菜...
10 持有筷子：2 等待　筷子：3
哲学家：2 ---吃菜...
9 持有筷子：1 等待　筷子：2
哲学家：1 ---吃菜...
11 持有筷子：3 等待　筷子：4
哲学家：3 ---吃菜...
12 持有筷子：4 等待　筷子：0
哲学家：4 ---吃菜...
8 持有筷子：0 等待　筷子：1

```
哲学家：0 ---吃菜...
10 持有筷子：2 等待 筷子：3
11 持有筷子：3 等待 筷子：4
8 持有筷子：0 等待 筷子：1
12 持有筷子：4 等待 筷子：0
9 持有筷子：1 等待 筷子：2
```

因为死锁如果发生的话，必须要同时满足前面描述的四个必要条件。所以如果想防止死锁，就需要考虑破坏其中任意一个条件。在本案例中每个哲学家都是先拿起左手筷子，然后再取右手筷子，如果把某个哲学家取筷子的顺序调整，就永远不会发生死锁了。

```
philosophers[philosophers.length-1] = new Philosopher(left,first);
```

把上面这句代码调整如下，就不会发生死锁了：

```
philosophers[i] = new Philosopher(first,left);    //不会死锁
```

3.9 本章习题

（1）对于如下代码，描述正确的是（　　）。

```java
public class Test {
    Object object = new Object();
    public void m1() {
        synchronized(object) {
            try {
                object.wait();
            } catch (Exception e) {
                e.printStackTrace();
            }
        }
    }
}
```

A. 调用方法 m1()，执行 object 对象的 wait()方法，会使当前线程进入阻塞等待状态
B. 调用方法 m1()，执行 object 对象的 wait()方法，会使当前线程进入 WAITING 等待状态
C. 调用 object 对象的 wait()方法前，必须要获得 object 对象的监视器锁，否则会抛出异常
D. 调用 object 对象的 wait()方法，与锁无关

（2）唤醒处于 WAITING 状态的线程，可以调用什么方法？（　　）

A. 调用 Object 对象的 notify()

B. 调用 Thread 的 stop()
C. 调用 Object 对象的 notifyAll()
D. 调用 Thread 的 interrupt()

（3）在线程 A 中，调用 Thread 对象的 join()方法，会使线程 A 进入什么状态？（　　）

A. WAITING 等待状态
B. BLOCKED 阻塞状态
C. TIMED_WAITING 延时等待状态
D. TERMINATED 终结状态

（4）调用线程的 interrupt()方法，会出现什么结果？（　　）

A. 不管这个线程处于什么状态，都会被中断结束
B. interrupt()方法并不能马上停止线程的运行，它只是给线程设置一个中断状态值，这相当于一个停止线程运行的建议，线程是否能够停止，由操作系统和 CPU 决定
C. 如果线程处于 BLOCKED 或 WAITING 状态，调用 interrupt()方法，那么它的中断状态将被清除，并且将收到一个 InterruptedException 异常
D. 调用 interrupt()方法与调用 stop()方法的效果相同

（5）如下的哪个条件，不属于死锁的必要条件？（　　）

A. 多线程使用的资源中至少有一个不能共享，即资源互斥
B. 线程持有资源并等待另一个资源
C. 高优先级的线程可以抢占低优先级线程的资源
D. 线程循环等待资源：一直不退出

第 4 章 线程池入门

线程池与数据库连接池非常相似,目的是提高服务器的响应能力。线程池可以设置一定数量的空闲线程,这些线程即使在没有任务时仍然不会释放。线程池也可以设置最大线程数,防止任务过多,压垮服务器。

4.1 ThreadPoolExecutor

ThreadPoolExecutor 是应用最广的底层线程池类,它实现了 Executor 和 ExecutorService 接口。在如下类的描述中,列举了 ThreadPoolExecutor 的构造函数和最常用的几个方法。

```java
public class ThreadPoolExecutor extends AbstractExecutorService{
    public ThreadPoolExecutor(int corePoolSize,
        int maximumPoolSize,
        long keepAliveTime,
        TimeUnit unit,
        BlockingQueue<Runnable> workQueue) {}
    @Override
    public void execute(Runnable command) {    }
    @Override
    public void shutdown() {}
    @Override
    public List<Runnable> shutdownNow() {}
    @Override
    public boolean isShutdown() {}
    @Override
    public boolean isTerminated() {    }
    @Override
    public boolean awaitTermination(long timeout, TimeUnit unit)
            throws InterruptedException {}
    ...
}
```

4.1.1 创建线程池

下面创建一个线程池,通过调整线程池构造函数的参数来了解线程池的运行特性。把核心线程数设置为 3,最大线程数设置为 8,阻塞队列的容量设置为 5。

(1) 当要执行的任务数小于核心线程数时,直接启动与任务数相同的工作线程。

```java
public static void main(String[] args) {
    BlockingQueue<Runnable> bq = new LinkedBlockingQueue<>(5);
    ThreadPoolExecutor pool = new
            ThreadPoolExecutor(3,8,2000,TimeUnit.MILLISECONDS,bq);
    for(int i=0;i<2;i++) {
        pool.execute(new Runnable() {
            public void run() {
                System.out.println(Thread.currentThread().getId()
                                    + " is running...");
                try {
                    Thread.sleep(800);
                } catch (Exception e) { }
            }
        });
    }
    pool.shutdown();
}
```

任务数量设置为 2,运行结果如下:

```
9 is running...
8 is running...
```

(2) 当任务数量大于核心线程数时,超过核心线程数的任务会自动加入阻塞队列中,直到把阻塞队列装满。

调整任务数量为 5:for(int i=0;i<5;i++) {...},观察程序运行结果如下。前面的 3 个任务启动了 3 个线程并加入线程池,后面的两个任务加入阻塞队列,等待前面的 3 个任务执行完毕。等前面 3 个任务完成后,程序会从阻塞队列中取出后面两个任务,然后仍然使用核心线程执行。因此会发现执行最后两个任务的线程号与前面的相同。

```
8 is running...
10 is running...
9 is running...
10 is running...
8 is running...
```

(3) 继续增加任务数量为 10:for(int i=0;i<10;i++) {...},观察程序的运行结果如下。仔

细观察会发现一共启动了 5 个线程。为什么线程池中的工作线程为 5 呢？

　　原因如下：核心线程数为 3，因此前面的 3 个任务会启动 3 个工作线程。阻塞队列数量为 5，因此第 4、5、6、7、8 这 5 个任务会自动加入阻塞队列。这时阻塞队列已满，第 9、10 两个任务会再启动两个新线程。注意：现在的工作线程数量一共为 5，小于线程池设置的最大线程数 8。

```
9 is running...
11 is running...
12 is running...
8 is running...
10 is running...
9 is running...
11 is running...
10 is running...
12 is running...
8 is running...
```

　　（4）继续增加任务数量为 13：for(int i=0;i<13;i++) {...}，观察程序的运行结果。程序运行步骤如下：先启动 3 个核心线程，然后 5 个任务进入阻塞队列，剩下的 5 个任务再次启动新的线程。工作线程数一共为 8 个，与线程池设置的最大线程数相符。

```
9 is running...
8 is running...
10 is running...
11 is running...
15 is running...
12 is running...
14 is running...
13 is running...
10 is running...
9 is running...
8 is running...
12 is running...
11 is running...
```

　　（5）继续增加任务数量为 15：for(int i=0;i<15;i++) {...}，观察程序的运行结果可以发现，当任务数大于"最大线程数+阻塞队列容量"时，会抛出 RejectedExecutionException（拒绝执行任务）异常。当前线程池的设置参数，最大容量是 8+5=13，当任务数超过 13 时，都会被拒绝。

```
8 is running...
10 is running...
12 is running...
```

```
9 is running...
13 is running...
11 is running...
14 is running...
15 is running...
Exception in thread "main" java.util.concurrent.RejectedExecutionException:
Task com.icss.pool.MyPool$1@1c7c054 rejected from...
12 is running...
9 is running...
8 is running...
13 is running...
10 is running...
```

（6）任务数设置为 13，观察线程池中的线程数量。

```
public static void main(String[] args) {
    BlockingQueue<Runnable> bq = new LinkedBlockingQueue<>(5);
    ThreadPoolExecutor pool = new
        ThreadPoolExecutor(3,8,2000,TimeUnit.MILLISECONDS,bq);
    for(int i=0;i<13;i++) {
        ...
    }
    System.out.println("池中线程数: " + pool.getPoolSize());
    try {
        Thread.sleep(5000);
    } catch (Exception e) {
    }
    System.out.println("池中线程数: " + pool.getPoolSize());
    pool.shutdown();
}
```

程序运行结果如下，线程池中的线程数最大为 8，当所有的任务都运行完成后，非核心线程数会在 2 秒后释放（参见前面 ThreadPoolExecutor 构造函数中的 keepAliveTime 参数设置），而核心线程数即使所有任务都已经完成也不会被释放。

```
8 is running...
10 is running...
12 is running...
9 is running...
11 is running...
13 is running...
池中线程数: 8
14 is running...
15 is running...
```

```
8 is running...
14 is running...
9 is running...
12 is running...
13 is running...
```
池中线程数：3

4.1.2 关闭线程池

调用 ThreadPoolExecutor 的 shutdown()方法或 shutdownNow()方法，可以关闭线程池。

```java
public class ThreadPoolExecutor extends AbstractExecutorService{
    @Override
    public void shutdown() {}
    @Override
    public List<Runnable> shutdownNow() {}
    @Override
    public boolean isShutdown() {}
}
```

在 4.1.1 节的代码上稍作变动，调用 shutdown()方法 3 秒后，再次读取线程池中的线程数量。

```java
public static void main(String[] args) {
    BlockingQueue<Runnable> bq = new LinkedBlockingQueue<>(5);
    ThreadPoolExecutor pool = new
        ThreadPoolExecutor(3,8,2000,TimeUnit.MILLISECONDS,bq);
    for(int i=0;i<13;i++) {
        ...
    }
    System.out.println("池中线程数：" + pool.getPoolSize());
    pool.shutdown();
    try {
        Thread.sleep(3000);
    } catch (Exception e) { }
    System.out.println("池中线程数：" + pool.getPoolSize());
}
```

测试结果如下，调用 shutdown()方法 3 秒后，再次读取线程池中的线程数量，发现其为 0，这表明核心线程也被停止运行了。调用 shutdown()方法后，原来提交的任务会被有序执行，但是不会再接受新的任务。

```
8 is running...
10 is running...
```

```
9 is running...
11 is running...
12 is running...
13 is running...
池中线程数：8
14 is running...
15 is running...
8 is running...
9 is running...
10 is running...
11 is running...
12 is running...
池中线程数：0
```

修改上面的测试代码，shutdown()方法修改为 shutdownNow ()方法，再次测试。

```java
public static void main(String[] args) {
    BlockingQueue<Runnable> bq = new LinkedBlockingQueue<>(5);
    ThreadPoolExecutor pool = new
        ThreadPoolExecutor(3,8,2000,TimeUnit.MILLISECONDS,bq);
    for(int i=0;i<13;i++) {
        ...
    }
    System.out.println("池中线程数：" + pool.getPoolSize());
    List<Runnable> undo = pool.shutdownNow();
    System.out.println("未执行的任务：" + undo.size());
}
```

测试结果如下，shutdownNow()与 shutdown()的主要区别是：shutdownNow()可以把已提交但是未执行的任务主动取消，并返回未执行的任务列表。

```
8 is running...
9 is running...
10 is running...
11 is running...
12 is running...
13 is running...
14 is running...
池中线程数：8
未执行的任务：5
15 is running...
```

4.2 Executor 接口

ThreadPoolExecutor 实现了 java.util.concurrent.Executor 接口。

```
public interface Executor {
    void execute(Runnable command);
}
```

Executor 接口中，只有一个 execute()方法。它表明在将来某个时刻，执行一个给定的任务。这个任务可以在一个新线程中或线程池中执行，也可以在调用 execute()方法的这个线程中执行。

执行器 Executor 把任务的提交和任务的运行进行了有机分解，从而实现了解耦。通常无须显示创建线程，如 new Thread(new(RunnableTask())).start()，而是按照如下方式创建线程并执行任务：

```
Executor executor = anExecutor;
executor.execute(new RunnableTask1());
executor.execute(new RunnableTask2());
...
```

Executor 接口并不严格要求任务执行是异步的，最简单的情况是，执行程序可以立即在调用者的线程中运行提交的任务。

Executor 接口是线程池技术的顶层接口，其他业务类和接口基本都继承或实现了 Executor 接口。类图如图 4-1 所示。

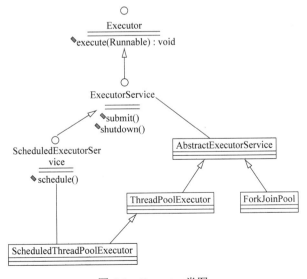

图 4-1 Executor 类图

ThreadPoolExecutor 的 execute()方法源代码如下，这个任务可以在一个新的线程中执行，也可以使用线程池中的已有线程执行。如果执行器已经被 shutdown，或达到了执行器的容量上限，任务将不能被提交，这会导致抛出 RejectedExecutionException 异常。

```java
public void execute(Runnable command) {
    if (command == null)
        throw new NullPointerException();
    int c = ctl.get();
    if (workerCountOf(c) < corePoolSize) {
        if (addWorker(command, true))
            return;
        c = ctl.get();
    }
    if (isRunning(c) && workQueue.offer(command)) {
        int recheck = ctl.get();
        if (! isRunning(recheck) && remove(command))
            reject(command);
        else if (workerCountOf(recheck) == 0)
            addWorker(null, false);
    }
    else if (!addWorker(command, false))
        reject(command);
}
```

任务执行过程主要分为如下几步：

（1）任务入参对象 command 为空，返回空指针异常。

（2）从 AtomicInteger 中获取原子数 c，调用 workerCountOf()方法，返回执行器的工作线程数。

（3）如果当前执行器工作线程数小于 corePoolSize 数量，则尝试创建一个新线程执行任务 command，并把新线程加入工作线程队列中。

（4）如果当前执行器工作线程数大于或等于 corePoolSize 数量，则尝试把任务 command 加入阻塞队列中。

（5）如果任务不能加入阻塞队列中，而且工作线程数小于 maximumPoolSize 数量，则尝试启动一个新的线程执行任务 command。

（6）如果执行器正在 shutdown，或工作线程数已达到 maximumPoolSize 数量，则任务 command 无法处理，会抛出 RejectedExecutionException 异常。

4.3　ExecutorService 接口

java.util.concurrent 包中的 ExecutorService 接口继承了 Executor 接口。Executor 接口只提供了任务的执行方法，为了解决执行服务对象的生命周期问题，ExecutorService 添加了一些用于生命周期管理的方法，如终止任务、提交任务、跟踪任务返回结果等。

```java
public interface ExecutorService extends Executor {
    void shutdown();
       List<Runnable> shutdownNow();
    boolean isShutdown();
    boolean isTerminated();
    <T> Future<T> submit(Callable<T> task);
}
```

ExecutorService 认为服务对象的生命期有 3 种状态：运行、关闭和终止。调用 shutdown() 方法可以有序关闭任务，调用 sutdownNow() 方法可以强制关闭未执行的任务，一旦所有的任务都执行完毕，ExecutorService 会转入终止状态。

下面比较一下 submit() 与 execute() 执行任务时的区别。submit() 方法有返回值 Future，而 execute() 方法没有任何返回信息。

```java
public interface ExecutorService extends Executor {
    <T> Future<T> submit(Callable<T> task);
}
public interface Executor {
    void execute(Runnable command);
}
```

4.3.1　Callable 返回任务执行结果

Callable 和 Runnable 都是执行任务的接口，但是 Runnable 接口中的 run() 方法无返回结果，而 Callable 的 call() 方法可以返回执行结果。

同时 Callable 接口的 call() 方法允许抛出异常，而 Runnable 接口中的 run() 方法异常只能在内部消化(日志)，不能抛给线程的调用者。

```java
public interface Callable<V> {
    V call() throws Exception;
}
```

如下示例，调用 ExecutorService 接口的 submit() 方法，调用线程可以接收异步线程的返回值。

（1）新建任务类 MyTask，实现 Callable 接口，call() 方法的返回数据为 String 类型。

```java
public class MyTask implements Callable<String> {
    private int id;
    public MyTask(int id) {
        this.id = id;
    }
```

```
    public String call() throws Exception {
        String result = "线程" + Thread.currentThread().getId() + ",id=" + id;
        return result;
    }
}
```

（2）新建线程池，调用 submit()方法，并使用集合接收所有返回结果。

```
public static void main(String[] args) {
    ExecutorServicepool = new ThreadPoolExecutor(3,8,2000,
    TimeUnit.MILLISECONDS, new LinkedBlockingQueue<Runnable>(5));
    List<Future<String>> results = new ArrayList<>();
    for(int i=0;i<10;i++) {
        Future<String> f = pool.submit(new MyTask(i));
        results.add(f);
        System.out.println("任务：" + i + "完成");
    }
    pool.shutdown();
}
```

（3）输出 call()方法的返回结果。注意：Future 的 get()方法需要阻塞等待任务完成，因此不能在线程池调用 submit()后马上调用 f.get()，否则会阻塞后面任务的执行。通常先把所有 call()方法的结果添加到集合后，再输出，这样不会影响线程池的并发性能。

```
public static void main(String[] args) {
    ...
    pool.shutdown();
    for(Future<String> f : results) {
        try {
            System.out.println(f.get());
            Thread.sleep(800);
        } catch (Exception e) {
        }
    }
}
```

输出结果如下，前面的异步任务是并行输出的，后面的返回结果是串行输出的：

任务：0 完成
任务：1 完成
任务：2 完成
任务：3 完成
任务：4 完成
任务：5 完成

任务：6 完成
任务：7 完成
任务：8 完成
任务：9 完成
线程 8,id=0
线程 9,id=1
线程 10,id=2
线程 8,id=3
线程 9,id=4
线程 9,id=5
线程 8,id=6
线程 9,id=7
线程 10,id=8
线程 10,id=9

4.3.2 shutdown 与 shutdownNow

调用 ExecutorService 接口的 shutdown()方法，启动有序关闭，其中先前提交的任务将仍然被执行，但是不会再接受任何新的任务。调用 shutdownNow()方法，会尝试停止所有任务，包括正在执行的任务和等待处理的任务，这个方法会返回未执行的任务列表。关于线程池关闭的测试，参见 4.1.2 节。

```
public interface ExecutorService extends Executor {
    List<Runnable> shutdownNow();
    void shutdown();
}
```

4.4 Executors 工具箱

Executors 工具箱为 Executor、ExecutorService、ScheduledExecutorService、ThreadFactory 和 Callable 等提供了工厂方法或工具方法。

该类支持以下几种方法：

（1）创建并返回一个 ExecutorService 对象，推荐了几种常用线程池的配置方式。
（2）创建并返回一个 ScheduledExecutorService 对象，设置了常用配置方式。
（3）创建并返回包装的 ExecutorService 对象，通过设置特定的方法禁用重新配置。
（4）创建并返回 ThreadFactory 对象，并设置创建的线程为已知状态。
（5）创建并返回一个 Callable 对象，这样在需要的执行方法中可以使用 Callable。

4.4.1 newCachedThreadPool

当需要处理大量的短时任务时,可以使用 Executors.newCachedThreadPool()创建线程池。参见 newCachedThreadPool()的源代码配置,可以更好地理解这个线程池的特性。

```
public class Executors {
    public static ExecutorService newCachedThreadPool() {
        return new ThreadPoolExecutor(0, Integer.MAX_VALUE,
                                      60L, TimeUnit.SECONDS,
                                      new SynchronousQueue<Runnable>());
    }
}
```

参考上述源代码,可以简单总结这个线程池的特点如下:

(1) ThreadPoolExecutor 的第一个参数 corePoolSize 为 0,即表示所有的任务都会进入阻塞队列。

(2) 这里使用的 SynchronousQueue 是一个非常独特的阻塞队列,它的内部容积为 0,只有工作线程准备好了做移除操作时,任务的插入操作才能成功,即任务插入与线程移除必须同时进行。

(3) ThreadPoolExecutor 的第二个参数 maximumPoolSize 为 Integer.MAX_VALUE,这是一个非常大的数字,即这个线程池理论上对线程数量没有限制,可以满足大量并发请求。

(4) ThreadPoolExecutor 的第三个参数 keepAliveTime 为 60 秒,即任务完成后,工作线程要 60 秒后才释放。这意味着大量的并发请求,可能创建新线程来处理,也可能使用池子中已完成任务但是未释放的线程来执行任务。

(5) 从上面的 ThreadPoolExecutor 参数配置,可以看出 newCachedThreadPool()创建的线程池,主要适合大量的短时任务处理。如果是长时任务,而且访问量非常大,则服务器无法承担这么大的负载,是一定会崩溃的。这个线程池未对任务数量做约束,而且使用 SynchronousQueue 减少入队与出队时间,就是为了快速满足高并发请求,即大量的并发请求无须等待,即可获得服务器的线程处理。当然,前提条件就是这些任务都是短时任务,这样服务器中的线程数量才是可控的。

案例描述:Web 服务器是一个多线程并发环境,它允许大量客户端通过 HTTP 同时访问 Web 服务器资源,如 Tomcat 就是应用最为广泛的 Web 服务器。客户端的 HTTP 请求,都是短时任务,Web 服务器需要快速响应客户端请求,返回给客户端需要的资源后,马上关闭客户端连接。这种场景使用 newCachedThreadPool()创建线程池,响应客户端请求是非常合适的。

下面模拟客户端并发请求 Web 服务器资源和 Web 服务器处理并发请求。

代码操作步骤如下。

(1) 模拟 Web 服务器,在 start()方法中启动监听器,同时调用 Executors 类的

newCachedThreadPool()创建线程池。

```java
public class WebServer {
    public void start() {
        try {
            ServerSocket server = new ServerSocket(9000);
            ExecutorService pool = Executors.newCachedThreadPool();
            System.out.println("Web 服务器启动,端口号 9000...");
            while (!pool.isShutdown()) {
                final Socket socket = server.accept();
                pool.execute(handleRequest(socket));
            }
        } catch (Exception e) {
            e.printStackTrace();
        }
    }
    ...
}
```

（2）Web 服务器每接收到一个客户端 HTTP 请求，都会创建一个新的 socket 对象，然后调用线程池的 execute()方法，让线程池处理客户端请求。

```java
public Runnable handleRequest(Socket socket) {
    Runnable runnable = new Runnable() {
        public void run() {
            try {
                DataInputStream dis = new DataInputStream(
                        new BufferedInputStream(socket.getInputStream()));
                byte[] buffer = new byte[1024];
                ByteArrayOutputStream bos = new ByteArrayOutputStream();
                int len = 0;
                while ((len = dis.read(buffer)) != -1) {
                    bos.write(buffer,0, len);
                }
                String reqMsg = new String(bos.toByteArray(),"iso-8859-1");
                System.out.println(Thread.currentThread().getId()
                                    + "收到:" + reqMsg);
                bos.close();
                dis.close();
                socket.close();
            } catch (Exception e) {
                e.printStackTrace();
            }
        }
```

```
    };
    return runnable;
}
```

（3）在主函数中启动 Web 服务器，随时接收客户端请求。

```
public static void main(String[] args) {
    new Thread(new Runnable() {
        public void run() {
            new WebServer().start();
        }
    }).start();
}
```

（4）模拟客户端大量 HTTP 请求，因为是并发请求，所以也使用 Executors 的 newCachedThreadPool()执行任务。不能在 for 循环中使用 new Thread()创建线程执行任务，因为那样就是顺序的请求，不是并发请求了。

```
public static void main(String[] args) {
    //模拟客户端大量并发请求
    ExecutorService pool = Executors.newCachedThreadPool();
    for (int i = 0; i < 50; i++) {
        pool.execute(new Runnable() {
            public void run() {
                try {
                    final Socket socket = new Socket("localhost", 9000);
                    DataOutputStream dos = new DataOutputStream(
                            new BufferedOutputStream(socket.getOutputStream()));
                    String msg = Thread.currentThread().getId() + ",hello";
                    dos.write(msg.getBytes("iso-8859-1"));
                    System.out.println("发送: " + msg);
                    dos.flush();
                    socket.shutdownOutput();
                    dos.close();
                    socket.close();
                } catch (Exception e) {
                    e.printStackTrace();
                }
            }
        });
    }
}
```

（5）客户端并发请求信息如下：

发送：44,hello
发送：42,hello
发送：9,hello
发送：51,hello
发送：12,hello
发送：39,hello
发送：37,hello
发送：14,hello
发送：11,hello
发送：32,hello
发送：31,hello
发送：53,hello
发送：40,hello
发送：35,hello
发送：10,hello
发送：13,hello
发送：34,hello
发送：41,hello
发送：15,hello
发送：21,hello
发送：47,hello
发送：23,hello
发送：30,hello
发送：22,hello
发送：20,hello
发送：27,hello
发送：54,hello
发送：48,hello
发送：33,hello
发送：56,hello
发送：19,hello
发送：18,hello
发送：17,hello
发送：25,hello
发送：43,hello
发送：26,hello
发送：50,hello
发送：45,hello
发送：55,hello
发送：29,hello
发送：46,hello
发送：36,hello
发送：28,hello

发送：24,hello
发送：16,hello
发送：57,hello
发送：8,hello
发送：49,hello
发送：38,hello
发送：52,hello

（6）Web 服务器接收并处理请求信息如下：

Web 服务器启动,端口号 9000...
10 收到:30,hello
12 收到:44,hello
11 收到:47,hello
11 收到:16,hello
11 收到:14,hello
11 收到:41,hello
15 收到:48,hello
11 收到:29,hello
12 收到:54,hello
11 收到:21,hello
15 收到:55,hello
19 收到:56,hello
11 收到:10,hello
15 收到:17,hello
11 收到:32,hello
12 收到:28,hello
12 收到:40,hello
15 收到:49,hello
12 收到:8,hello
15 收到:50,hello
10 收到:52,hello
11 收到:43,hello
16 收到:13,hello
23 收到:37,hello
14 收到:57,hello
17 收到:12,hello
31 收到:42,hello
18 收到:24,hello
13 收到:9,hello
20 收到:23,hello
38 收到:46,hello
21 收到:53,hello
24 收到:35,hello

```
28 收到:18,hello
25 收到:11,hello
32 收到:31,hello
36 收到:33,hello
29 收到:36,hello
40 收到:51,hello
37 收到:26,hello
33 收到:39,hello
27 收到:34,hello
35 收到:38,hello
22 收到:45,hello
26 收到:15,hello
30 收到:25,hello
42 收到:19,hello
41 收到:22,hello
39 收到:20,hello
34 收到:27,hello
```

4.4.2 newFixedThreadPool

调用 Executors 的 newFixedThreadPool()方法，可以创建一个线程池，它的线程数量始终固定。当任务数量大于线程数量时，需要在阻塞队列中排队等待执行。其源代码定义如下：

```java
public class Executors {
    public static ExecutorService newFixedThreadPool(int nThreads) {
        return new ThreadPoolExecutor(nThreads, nThreads,
                                      0L, TimeUnit.MILLISECONDS,
                                      new LinkedBlockingQueue<Runnable>());
    }
}
```

参考上述源代码，可以简单总结这个线程池的特点如下：

（1）ThreadPoolExecutor 的第一个参数 corePoolSize 为入参 nThreads，即该线程池的固定线程数量就是 nThreads 个。

（2）这个线程池使用的阻塞队列为 LinkedBlockingQueue，这是一个无界阻塞队列，容量默认为 Integer.MAX_VALUE，非常大。这意味着，当任务数量大于核心线程数时，都会被装入 LinkedBlockingQueue 队列中，这里有充分的空间容纳待执行的任务。

（3）这个线程池的其他参数，如 maximumPoolSize=nThreads，基本没有什么用处。因为阻塞队列的空间足够大，线程数量无法超过 nThreads，maximumPoolSize 设置得再大也没用。

（4）keepAliveTime 用于表示空闲线程在回收前的等待时间，这个线程池不会出现空闲线程,因此 keepAliveTime=0 也是无用的设置。

案例场景描述:现在有 20 个学员要参加机动车驾驶证考试,由于场地限制,每次最多允许 3 个人进行考试,其他人排队等候。所有学员考试完成后,则考试结束。

实现步骤如下。

(1)新建学员类 Student。

```java
class Student {
    private String name;
    public String getName() {
        return name;
    }
    public Student(String name) {
        this.name = name;
    }
}
```

(2)在 3 个考场中对 20 个学员进行机动车驾驶证路考。

```java
public static void main(String[] args) {
    ExecutorService pool = Executors.newFixedThreadPool(3);
    for(int i=0;i<20;i++) {
        Student stu = new Student("s" + i);
        pool.execute(new Runnable() {
            public void run() {
                try {
                    System.out.println("学生" + stu.getName() + "开始考试,
                            考场:" + Thread.currentThread().getId());
                    Thread.sleep((int)(Math.random()*1000));
                    System.out.println("学生"+stu.getName()+"结束考试...");
                } catch (Exception e) {
                    e.printStackTrace();
                }
            } });  }
    pool.shutdown();
}
```

(3)考试模拟输出效果如下:

学生 s0 开始考试,考场:8
学生 s1 开始考试,考场:9
学生 s2 开始考试,考场:10
学生 s1 结束考试...
学生 s0 结束考试...
学生 s2 结束考试...
学生 s4 开始考试,考场:8

学生 s3 开始考试,考场:9
学生 s4 结束考试...
学生 s5 开始考试,考场:10
学生 s6 开始考试,考场:8
学生 s3 结束考试...
学生 s6 结束考试...
学生 s5 结束考试...
学生 s8 开始考试,考场:8
学生 s7 开始考试,考场:9
学生 s8 结束考试...
学生 s9 开始考试,考场:10
学生 s10 开始考试,考场:8
学生 s7 结束考试...
学生 s10 结束考试...
学生 s9 结束考试...
学生 s12 开始考试,考场:8
学生 s11 开始考试,考场:9
学生 s12 结束考试...
学生 s13 开始考试,考场:10
学生 s14 开始考试,考场:8
学生 s13 结束考试...
学生 s11 结束考试...
学生 s15 开始考试,考场:10
学生 s14 结束考试...
学生 s15 结束考试...
学生 s16 开始考试,考场:9
学生 s18 开始考试,考场:10
学生 s17 开始考试,考场:8
学生 s18 结束考试...
学生 s16 结束考试...
学生 s19 开始考试,考场:10
学生 s19 结束考试...
学生 s17 结束考试...

（4）如果某个考场在考试中出现异常，处理异常后，不会影响线程池中的固定线程个数，后面的任务会正常处理。

```
public static void main(String[] args) {
    ExecutorService pool = Executors.newFixedThreadPool(3);
    for(int i=0;i<20;i++) {
        Student stu = new Student("s" + i);
        pool.execute(new Runnable() {
            public void run() {
```

```java
            try {
                System.out.println("学生"+stu.getName()+"开始考试,考场:"
                        + Thread.currentThread().getId());
                int rand = (int)(Math.random()*10);
                if(rand > 8) {
                    throw new Exception("考场" +
                        Thread.currentThread().getId() + "异常");
                }
                Thread.sleep((int)(Math.random()*1000));
                System.out.println("学生" + stu.getName()+"结束考试...");
            } catch (Exception e) {
                e.printStackTrace();
            }
        } });
        pool.shutdown();
    }
```

（5）当考试中抛出异常后，后面的任务正常处理，程序的运行结果如下：

```
学生 s0 开始考试,考场:8
学生 s1 开始考试,考场:9
学生 s2 开始考试,考场:10
学生 s1 结束考试...
学生 s3 开始考试,考场:9
学生 s0 结束考试...
学生 s4 开始考试,考场:8
学生 s2 结束考试...
学生 s5 开始考试,考场:10
学生 s3 结束考试...
学生 s6 开始考试,考场:9
java.lang.Exception: 考场 9 异常
    at com.icss.pool.Exam$1.run(Exam.java:19)
    at java.lang.Thread.run(Thread.java:745)
学生 s7 开始考试,考场:9
学生 s4 结束考试...
学生 s8 开始考试,考场:8
学生 s5 结束考试...
学生 s9 开始考试,考场:10
学生 s7 结束考试...
学生 s10 开始考试,考场:9
学生 s8 结束考试...
学生 s11 开始考试,考场:8
学生 s11 结束考试...
学生 s12 开始考试,考场:8
```

```
学生 s10 结束考试...
java.lang.Exception: 考场 9 异常
    at com.icss.pool.Exam$1.run(Exam.java:19)
    at java.lang.Thread.run(Thread.java:745)
学生 s13 开始考试,考场:9
学生 s14 开始考试,考场:9
学生 s12 结束考试...
学生 s15 开始考试,考场:8
学生 s9 结束考试...
学生 s16 开始考试,考场:10
学生 s14 结束考试...
学生 s17 开始考试,考场:9
学生 s16 结束考试...
学生 s18 开始考试,考场:10
学生 s19 开始考试,考场:10
java.lang.Exception: 考场 10 异常
    at com.icss.pool.Exam$1.run(Exam.java:19)
    at java.lang.Thread.run(Thread.java:745)
学生 s17 结束考试...
学生 s15 结束考试...
学生 s19 结束考试...
```

4.4.3　newSingleThreadExecutor

调用 Executors 的 newSingleThreadExecutor()方法,可以创建一个线程池,它的线程数量始终固定为 1 个。当任务数量大于 1 时,需要在阻塞队列中排队等待执行。其源代码定义如下:

```java
public class Executors {
    public static ExecutorService newSingleThreadExecutor() {
        return new FinalizableDelegatedExecutorService
            (new ThreadPoolExecutor(1, 1,
                        0L, TimeUnit.MILLISECONDS,
                        new LinkedBlockingQueue<Runnable>()));
    }
}
```

参考源代码,从 ThreadPoolExecutor 的配置参数,可以很容易看出,这个线程池的配置信息就是 Executors.newFixedThreadPool()配置参数的特例。即当前线程池也是固定线程数的线程池,只是这个固定的线程数为 1,不能改变罢了。

使用单一固定的线程处理业务,在很多场景都非常有用,所以在工具箱 Executors 中单独做了设置,如 Socket 连接监听器就是典型的应用场景。

案例场景描述：解耦业务逻辑与日志操作，在高并发环境，大量的写日志操作会影响业务操作的性能（日志文件的 I/O 会成为性能瓶颈），因此可以把写日志的操作在独立的线程池中完成。本案例在高并发网站中可以借助消息中间件实现，此处用线程池模拟。

操作步骤如下：

（1）包装并配置 log4j，实现日志操作。

```java
import org.apache.log4j.Logger;
public class Log {
    public static Logger logger = Logger.getLogger(Log.class.getName());
}
```

（2）定义类 LogWriter，提取日志队列中的消息，写入日志文件中。注意：此处调用的 BlockingQueue 的 take() 方法，为阻塞等待任务到达后移除任务。当队列中没有消息时，take() 方法始终处于阻塞等待状态。

```java
public class LogWriter implements Runnable{
    public static final BlockingQueue<String> queue
                = new LinkedBlockingQueue<String>();
    public void run() {
        while(true) {
            try {
                String msg = LogWriter.queue.take();
                Log.logger.info(Thread.currentThread().getId() + "," + msg
                        + ",待处理: " + LogWriter.queue.size());
            } catch (Exception e) {
                e.printStackTrace();
            }
        }
    }
}
```

（3）模拟业务类的业务方法，业务操作完成后，日志信息存入消息队列中。

```java
class Service {
    public void m1() throws Exception {
        System.out.println(Thread.currentThread().getId() + ",m1 working");
        LogWriter.queue.put("m1 do in " + Thread.currentThread().getId());
    }
    public void m2() throws Exception {
        System.out.println(Thread.currentThread().getId() + ",m2 working");
        LogWriter.queue.put("m2 do in " + Thread.currentThread().getId());
    }
    public void m3() throws Exception{
        System.out.println(Thread.currentThread().getId() + ",m3 working");
```

```
            LogWriter.queue.put("m3 do in " + Thread.currentThread().getId());
        }
    }
```

（4）在主函数中，创建单一线程池，专门用于写日志。

```
public static void main(String[] args) {
    ExecutorService pool = Executors.newSingleThreadExecutor();
    pool.execute(new LogWriter());
    ...
}
```

（5）在主函数中，使用 3 个线程模拟业务操作的并发场景。

```
public static void main(String[] args) {
    ...
    Service s = new Service();
    ExecutorService worker = Executors.newFixedThreadPool(3);
    for(int i=0;i<50;i++) {
        worker.execute(new Runnable() {
            public void run() {
                int rand = (int)(Math.random()*10);
                try {
                    if(rand<3) {
                        s.m1();
                    }else if(rand>3 && rand<6) {
                        s.m2();
                    }else {
                        s.m3();
                    }
                     Thread.sleep(300);
                } catch (Exception e) {
                    e.printStackTrace();
                }
            }
        });
    }
    worker.shutdown();
}
```

（6）程序执行效果如下，业务行为与写日志完全分离开来，不会因为文件的 I/O 影响业务操作的并发性能。

```
9,m1 working
10,m1 working
```

```
11,m3 working
INFO - 8,m1 do in 9,待处理: 2
INFO - 8,m1 do in 10,待处理: 1
INFO - 8,m3 do in 11,待处理: 0
11,m3 working
10,m3 working
9,m2 working
INFO - 8,m3 do in 11,待处理: 0
INFO - 8,m3 do in 10,待处理: 0
INFO - 8,m2 do in 9,待处理: 0
11,m1 working
INFO - 8,m1 do in 11,待处理: 0
10,m3 working
9,m3 working
INFO - 8,m3 do in 10,待处理: 0
INFO - 8,m3 do in 9,待处理: 0
11,m3 working
10,m3 working
9,m3 working
INFO - 8,m3 do in 11,待处理: 0
INFO - 8,m3 do in 10,待处理: 1
INFO - 8,m3 do in 9,待处理: 0
```

（7）若写日志线程在运行过程出现异常，被捕获后，不会影响线程池的继续运行。参见如下代码，在 rand=5 时，LogWriter 中抛出异常并捕获异常。

```java
public void run() {
    while(true) {
        try {
            String msg = LogWriter.queue.take();
            Log.logger.info(Thread.currentThread().getId() + "," + msg
                    + ",待处理: " + LogWriter.queue.size());
            int rand = (int)(Math.random()*10);
            if(rand == 5) {
                throw new RuntimeException("线程池异常测试...");
            }
        } catch (Exception e) {
            e.printStackTrace();
        }
    }
}
```

程序运行结果如下，抛出异步并处理后，程序运行正常（如果抛出异常后，未捕获处理，则线程池无法再继续工作)。

```
10,m3 working
11,m1 working
9,m3 working
INFO - 8,m3 do in 10,待处理: 2
INFO - 8,m1 do in 11,待处理: 1
INFO - 8,m3 do in 9,待处理: 0
java.lang.RuntimeException: 线程池异常测试...
    at com.icss.pool.LogWriter.run(LogWriter.java:21)
9,m2 working
10,m1 working
11,m3 working
INFO - 8,m2 do in 9,待处理: 0
INFO - 8,m1 do in 10,待处理: 1
INFO - 8,m3 do in 11,待处理: 0
java.lang.RuntimeException: 线程池异常测试...
    at com.icss.pool.LogWriter.run(LogWriter.java:21)
11,m3 working
10,m3 working
INFO - 8,m3 do in 11,待处理: 1
INFO - 8,m3 do in 10,待处理: 0
9,m3 working
INFO - 8,m3 do in 9,待处理: 0
java.lang.RuntimeException: 线程池异常测试...
    at com.icss.pool.LogWriter.run(LogWriter.java:21)
11,m3 working
10,m3 working
INFO - 8,m3 do in 10,待处理: 1
9,m3 working
INFO - 8,m3 do in 11,待处理: 0
INFO - 8,m3 do in 9,待处理: 0
```

4.4.4　newScheduledThreadPool

Executors 中的 newScheduledThreadPool()方法，用于创建一个指定大小 corePoolSize 的线程池，支持在给定的延时之后执行或定期执行任务。参见源代码如下：

```java
public class Executors {
    public static ScheduledExecutorService newScheduledThreadPool(
                    int corePoolSize) {
            return new ScheduledThreadPoolExecutor(corePoolSize);
    }
```

```java
}
public interface ScheduledExecutorService {
    public ScheduledFuture<?> schedule(Runnable command,
                                       long delay,
                                       TimeUnit unit);
    public <V> ScheduledFuture<V> schedule(Callable<V> callable,
                                           long delay, TimeUnit unit);
}
    public ScheduledFuture<?> scheduleAtFixedRate(
                                       Runnable command,
                                       long initialDelay,
                                       long period,
                                       TimeUnit unit);
}
```

ScheduledExecutorService 接口提供了各种延时任务和定时任务的执行，并返回可用于取消或检查执行任务的对象。

如下示例，演示了延时 5 秒后执行任务的线程池：

```java
public static void main(String[] args) {
    ScheduledExecutorService pool = Executors.newScheduledThreadPool(3);
    for(int i=0;i<10;i++) {
        pool.schedule(new Runnable() {
            public void run() {
                System.out.println(Thread.currentThread().getId() + " running");
            }
        }, 5, TimeUnit.SECONDS);
    }
    pool.shutdown();
}
```

程序在主函数启动 5 秒后执行任务，程序输出结果如下：

```
9 running
10 running
9 running
10 running
9 running
10 running
9 running
10 running
8 running
9 running
```

ScheuledThreadPoolExecutor 是 ScheduledExecutorService 接口的实现类，它将 Runnable 对象包装成了一个 ScheduledFutureTask，ScheduledFutureTask 是 ScheuledThreadPoolExecutor 中的内部类，它继承了 FutureTask 类并且实现了 RunnableScheduledFuture 接口。因此它拥有了取消异步计算和检索计算结果的能力。这看起来有些类似于 Timer 类，Timer 工具管理任务的延迟执行以及周期执行。但是 Timer 类存在一些缺陷。Timer 类只能创建唯一的线程来执行所有 Timer 任务。如果一个 Timer 任务的执行很耗时，会导致其他 TimerTask 的时效准确性出问题。

ScheuledThreadPoolExecutor 解决了这个缺陷，它可以提供多个线程来执行延迟、周期性的任务。Timer 的另一个问题在于，如果 TimerTask 抛出未检查的异常，Timer 将会产生无法预料的行为。Timer 线程并不捕获异常，所以 TimerTask 抛出未检查的异常将会终止 Timer 线程。在这种情况下，Timer 不会重新恢复线程的执行了。也就是说已经被安排但是尚未被执行的 TimerTask 永远也不会执行了。

DelayedWorkQueue 是 ScheduledThreadPoolExecutor 类中定义的静态内部类，DelayedWorkQueue 实现了 BlockingQueue 接口，所以它是一个阻塞队列。

```java
public class ScheduledThreadPoolExecutor {
    static class DelayedWorkQueue extends AbstractQueue<Runnable>
                                  implements BlockingQueue<Runnable> {}
}
```

参考 DelayedWorkQueue 源代码可知，DelayedWorkQueue 队列中的元素是用数组进行存储的，通过二叉树算法实现快速插入和删除，从根节点到子叶节点的每条路径都是降序的，所以它也叫优先级队列。它保证添加到队列中的任务，会按照任务的延时时间进行排序，延时时间少的任务首先被获取。

4.4.5 newWorkStealingPool

newWorkStealingPool 将所有可用的 CPU 处理器作为其目标，并行创建一个工作窃取线程池。这个线程池与前面的几个不同，它不是基于 ThreadPoolExecutor，而是基于 ForkJoinPool 做的优化配置。

```java
public class Executors {
    public static ExecutorService newWorkStealingPool() {
        return new ForkJoinPool
            (Runtime.getRuntime().availableProcessors(),
            ForkJoinPool.defaultForkJoinWorkerThreadFactory,
            null, true);
    }
}
```

1. ForkJoinPool

Fork 的英文单词是分叉、分支的意思,而 join 是连接、合并的意思。ForkJoinPool 从字面意思可以理解为:任务分解执行,然后把执行结果合并的线程池。

```
public class ForkJoinPool
                extends AbstractExecutorService {}
public abstract class AbstractExecutorService
                implements ExecutorService {}
```

ForkJoinPool 同 ThreadPoolExecutor 一样,实现了 Executor 和 ExecutorService 接口。ForkJoinPool 使用了一个无限队列来保存需要执行的任务,而线程的数量则是通过构造函数传入,如果没有向构造函数中传入希望的线程数量,那么当前计算机可用的处理器数量会被设置为默认值。

ForkJoinPool 充当 fork/join 框架里面的管理者,最原始的任务都要交给它才能处理。ForkJoinPool 负责控制整个 fork/join 有多少个工作线程,线程的创建与激活都是由它来掌控。ForkJoinPool 还负责 workQueue 队列的创建和分配,每创建一个工作线程,ForkJoinPool 都负责分配相应的 workQueue。然后 ForkJoinPool 把任务都交给工作线程去处理,ForkJoinPool 可以说是整个 fork/join 的容器。

ForkJoinPool 的优势是什么?ForkJoinPool 的优势在于,可以充分利用多处理器的优势,把一个任务拆分成多个比较小的任务,把这些小任务负载到多个处理器上并行执行。当这些小任务执行完成后,再将这些执行结果合并起来,如递归运算、快速排序算法等。

图 4-2 展示了 ForkJoinPool 对待任务"分而治之"的处理思想。

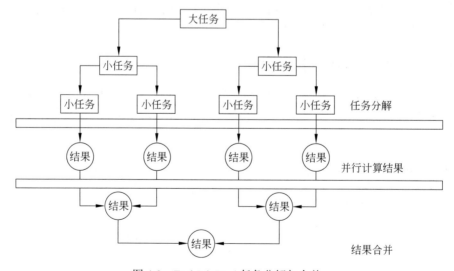

图 4-2　ForkJoinPool 任务分解与合并

2. ForkJoinTask

调用 ExecutorService 的 submit()方法，返回 ForkJoinTask 对象。

```java
public class ForkJoinPool {
    public <T> ForkJoinTask<T> submit(Callable<T> task) { }
    public ForkJoinTask<?> submit(Runnable task) { }
}
```

ForkJoinTask 实现了 Future 接口，它是一个线程实体，比普通线程的性能损耗低得多。

```java
public abstract class ForkJoinTask<V>
                implements Future<V>, Serializable {}
```

ForkJoinTask 同时也是一个轻量的 Future，使用时应避免较长时间阻塞和 I/O 处理，ForkJoinTask 是一个抽象类，在 Java8 中应用广泛。ForkJoinTask 中的 fork()方法用于任务分解，把一个任务分解为线程池 ForkJoinPool 执行的多个异步任务。join()方法用于任务合并。源代码参考如下：

```java
public abstract class ForkJoinTask<V>
                implements Future<V>, Serializable {
    public final ForkJoinTask<V> fork() {
        Thread t;
        if ((t = Thread.currentThread()) instanceof ForkJoinWorkerThread)
            ((ForkJoinWorkerThread)t).workQueue.push(this);
        else
            ForkJoinPool.common.externalPush(this);
        return this;
    }
    public final V join() {
        int s;
        if ((s = doJoin() & DONE_MASK) != NORMAL)
            reportException(s);
        return getRawResult();
    }
}
```

ForkJoinTask 存在两个子类：RecursiveAction 和 RecursiveTask。它们之间的区别是：RecursiveAction 没有返回值，RecursiveTask 有返回值。子类重写 compute()方法，即可调用 fork()与 join()方法，进行异步任务分解与合并。

```java
public abstract class RecursiveTask<V>
                extends ForkJoinTask<V> {
    V result;           //存储计算结果
    protected abstract V compute();
```

```java
}
public abstract class RecursiveAction
                    extends ForkJoinTask<Void> {
    protected abstract void compute();
}
```

3. 案例：斐波那契数列求解

案例场景描述：斐波那契数列又被称为"兔子数列"，是以兔子繁殖为例引入的，如 0、1、1、2、3、5、8、13、21、34……，它的特点是从第 3 项开始，每一项都等于前两项之和。可以使用 $f[n]=f[n-1]+f[n-2]$ 表示。

使用单线程模式，用递归算法计算第 n 项的斐波那契数列值。操作步骤如下：

（1）构造函数，传入要计算第几项斐波那契数列值。

```java
public class Fibonac {
    final int n;
    Fibonac(int n) {
        this.n = n;
    }
    ...
}
```

（2）在 compute()方法中，递归调用 $f(n-1)$ 与 $f(n-2)$ 的斐波那契数列值。

```java
public class Fibonac {
    ...
    protected Integer compute() {
        System.out.println(Thread.currentThread().getId() + " doing...");
        if (n <= 1)
            return n;
        Fibonac f1 = new Fibonac(n - 1);
        Fibonac f2 = new Fibonac(n - 2);
        return f1.compute() + f2.compute();
    }
}
```

（3）由于单线程反复递归，当 $n=30$ 时，Intel i3 机器的计算已经非常费力。由于 n 每增加 1，时间都呈指数级增加，当 $n>100$ 时，运算时间就非常恐怖了。

```java
public static void main(String[] args) {
    Fibonac fb = new Fibonac(30);
    long start = new Date().getTime();
    int result = fb.compute();
    long span = new Date().getTime() - start;
    System.out.println(result + ",用时" + span);
```

}

运行结果如下：

```
...
1 doing..
1 doing..
1 doing..
1 doing..
1 doing..
1 doing..
832040,用时 27492
```

使用 RecursiveTask 可以把斐波那契数列的计算任务，由单线程分担给多个子线程，所有子任务的计算结果可以使用 join()方法合并。

（4）定义 RecursiveTask 类，继承 RecursiveTask 抽象类。

```java
public class FibonacFork extends RecursiveTask<Integer> {
    final int n;
    FibonacFork(int n) {
        this.n = n;
    }
}
```

（5）重写 RecursiveTask 的 compute()方法。调用 fork()方法进行任务分解，调用 join()方法进行任务合并。

```java
protected Integer compute() {
    System.out.println(Thread.currentThread().getId() + " doing...");
    if (n <= 1)
        return n;
    FibonacFork f1 = new FibonacFork(n - 1);
    FibonacFork f2 = new FibonacFork(n - 2);
    f1.fork();
    f2.fork();
    return f1.join() + f2.join();
}
```

（6）主函数的调用方式不变，仍然计算斐波那契数列的第 30 项结果。

```java
public static void main(String[] args) {
    FibonacFork fb = new FibonacFork(30);
    long start = new Date().getTime();
    int result = fb.compute();
    long span = new Date().getTime() - start;
```

```
        System.out.println(result + ",用时" + span);
    }
```

程序运行结果如下，不过令人尴尬的是，采用多线程异步模式后，计算用时并没有减少，反而增加了。这充分说明，只有在真正的多 CPU 环境下，才能发挥 ForkJoin 异步运算的优势。

```
...
17 doing...
17 doing...
13 doing...
13 doing...
13 doing...
12 doing...
12 doing...
12 doing...
12 doing...
12 doing...
16 doing...
12 doing...
12 doing...
10 doing...
10 doing...
10 doing...
14 doing...
14 doing...
14 doing...
832040,用时 37132
```

（7）使用 ForkJoinPool，仍然计算斐波那契数列的第 30 项结果。

```java
public static void main(String[] args) {
    ForkJoinPool pool = (ForkJoinPool) Executors.newWorkStealingPool();
    FibonacFork task = new FibonacFork(30);
    long start = new Date().getTime();
    Future<Integer> future = pool.submit(task);
    try {
        int result = future.get();
        long span = new Date().getTime() - start;
        System.out.println(result + ",用时" + span);
        pool.shutdown();
    } catch (Exception e) {
        e.printStackTrace();
    }
}
```

}

测试结果显示,在单 CPU 环境,步骤(7)与步骤(6)的测试结果接近,性能没有明显的提升。

(8) 前面使用递归算法,计算斐波那契数列的第 *n* 项的结果。计算机运算效率非常低,这是我们为了演示 fork/join 故意设计的案例。其实可以采用非常简单的算法,在极短的时间计算出上述结果。

```java
public static int fac(int n) {
    int fn_2 = 0;
    int fn_1 = 1;
    int fn = 1;
    for(int i=2;i<=n;i++) {
        fn = fn_1 + fn_2;
        fn_2 = fn_1;
        fn_1 = fn;
    }
    return fn;
}
public static void main(String[] args) {
    System.out.println(fac(30));
}
```

4. 案例:大数组分段排序

案例场景描述:对一个长度为 5000 的大数组进行数值排序,使数组当中的每一个值都按照递增顺序排列。因为我们只需数字排序,无须统计结果,所以在这个案例中可以使用 RecursiveAction 类。

操作步骤如下。

(1) 定义排序任务类 SortTask,继承 RecursiveAction 类。

```java
class SortTask extends RecursiveAction {
    final long[] array;
    final int lo, hi;
    SortTask(long[] array, int lo, int hi) {
        this.array = array;
        this.lo = lo;
        this.hi = hi;
    }
    protected void compute() {
        System.out.println(Thread.currentThread().getId() + " doing...");
        // 200 一段,用 Arrays.sort 排序。大于 200 的数组段,切分
        if (hi - lo < 200) {
```

```java
            Arrays.sort(array, lo, hi);
        } else {
            int mid = (lo + hi) >>> 1;
            //分解任务并执行
            invokeAll(new SortTask(array, lo, mid), new SortTask(array, mid, hi));
            merge(lo, mid, hi);
        }
    }
    // 使用中间数组，进行比较交换
    void merge(int lo, int mid, int hi) {
        long[] buf = Arrays.copyOfRange(array, lo, mid);
        for (int i = 0, j = lo, k = mid; i < buf.length; j++)
            array[j] = (k == hi || buf[i] < array[k]) ? buf[i++] : array[k++];
    }
}
```

（2）定义一个大数组，填充随机数。调用 RecursiveAction 对象进行任务分解排序。

```java
public static void main(String[] args) {
    long[] array = new long[8000];
    for (int i = 0; i < 8000; i++) {
        int rand = (int) (Math.random() * 30000);
        array[i] = rand;
    }
    SortTask task = new SortTask(array,0,array.length);
    task.compute();
    for (long i : array) {
        System.out.println(i);
    }
}
```

（3）程序运行结果如下（部分节选）：

```
1 doing..
1 doing..
8 doing..
8 doing..
8 doing..
9 doing..
1 doing..
9 doing..
10 doing..
8 doing..
10 doing..
```

```
9 doing..
1 doing..
...
...
29947
29948
29954
29955
29959
29963
29968
29971
29972
29977
29979
29987
29988
```

4.5 线程工厂与线程组

创建线程可以用 new Thread()或从线程池中获取，为了更灵活地创建线程，需要使用线程工厂。线程池底层代码创建线程时使用的就是线程工厂对象。

线程组是为了管理线程而采用的分组模式。

4.5.1 线程组

线程组代表一组线程。每个线程组还可以包含其他线程组。线程组会形成一棵树，除了根节点线程组之外，每个线程组都有一个父线程组。

```
public class ThreadGroup
        extends Object
        implements Thread.UncaughtExceptionHandler
    private final ThreadGroup parent;        //当前线程组的父
    String name;                             //线程组的名字
    int maxPriority;                         //线程组的优先级
    boolean daemon;                          //是否为 daemon
    Thread threads[];                        //线程组中包含一组其他线程
    ThreadGroup groups[];                    //当前线程组包含其他线程组
    public ThreadGroup(ThreadGroup parent, String name) {
        this.name = name;
        this.maxPriority = parent.maxPriority;
        this.daemon = parent.daemon;
```

```java
            this.parent = parent;
            parent.add(this);
        }
    public ThreadGroup(String name) {
            this(Thread.currentThread().getThreadGroup(), name);
        }
}
```

默认情况下，根节点线程组的名字为 system，应用程序主线程组的名字为 main，通过下面的代码，可以测试当前线程的所有父线程组信息。

```java
public static void printParentGroupInfo() {
    ThreadGroup current = Thread.currentThread().getThreadGroup();
    System.out.println(current.getName() + "," + current.getMaxPriority());
    for(;;) {
        ThreadGroup parent = current.getParent();
        if(parent != null) {
            System.out.println(parent.getName() + "," + parent.getMaxPriority());
            current = parent;
        }else {
            break;
        }
    }
}
public static void main(String[] args) {
    printParentGroupInfo();
}
```

程序运行结果如下：

```
main,10
system,10
```

4.5.2 线程与线程组

线程 Thread 与线程组 ThreadGroup 关系紧密，在 Thread 类的所有构造函数中，都要调用 init()方法，初始化方法的第一个参数就是线程组，如果线程组的参数为空，会默认使用当前线程的线程组对象（当前线程就是创建者线程），即新创建的线程默认使用创建者的线程组。

```java
public class Thread implements Runnable {
    private ThreadGroup group;
    private long tid;                  //线程id
```

```java
public Thread() {
    init(null, null, "Thread-" + nextThreadNum(), 0);
}
public Thread(Runnable target) {
    init(null, target, "Thread-" + nextThreadNum(), 0);
}
public Thread(Runnable target, AccessControlContext acc) {
    init(null, target, "Thread-" + nextThreadNum(), 0, acc, false);
}
public Thread(ThreadGroup group, Runnable target) {
    init(group, target, "Thread-" + nextThreadNum(), 0);
}
public Thread(String name) {
    init(null, null, name, 0);
}
private void init(ThreadGroup g, Runnable target, String name,
    long stackSize, AccessControlContext acc,
    boolean inheritThreadLocals) {
    this.name = name;
    Thread parent = currentThread();
    if (g == null) {
        g = parent.getThreadGroup();   //默认使用创建者的线程组
    }
    this.group = g;
    this.daemon = parent.isDaemon();
    this.priority = parent.getPriority();
    tid = nextThreadID();
}
}
```

测试线程与线程组的关系，操作步骤如下。

（1）新建线程 Thread 时，传入线程组对象（未调用线程的 start 方法）。

```
ThreadGroup group = new ThreadGroup("worker");
Thread t1 = new Thread(group,"myThread");
System.out.println(t1.getState());
System.out.println("t1 线程所属线程组：" + t1.getThreadGroup().getName());
System.out.println("主线程所属线程组："
        + Thread.currentThread().getThreadGroup().getName());
```

程序运行结果如下：

```
NEW
t1 线程所属线程组：worker
```

主线程所属线程组：main

（2）在步骤（1）的基础上，启动线程后再进行测试。

```
ThreadGroup group = new ThreadGroup("worker");
Thread t1 = new Thread(group,"myThread");
t1.start();
System.out.println(t1.getState());
System.out.println("t1 线程所属线程组: " + t1.getThreadGroup().getName());
System.out.println("主线程所属线程组: "
                + Thread.currentThread().getThreadGroup().getName());
```

程序运行结果与步骤（1）的结果相同：

```
RUNNABLE
t1 线程所属线程组：worker
主线程所属线程组：main
```

（3）在步骤（2）的基础上，增加延时后再次读取线程组信息。

```
ThreadGroup group = new ThreadGroup("worker");
Thread t1 = new Thread(group,"myThread");
t1.start();
try {
    Thread.sleep(500);
} catch (Exception e) {
}
System.out.println(t1.getState());
System.out.println("t1 线程所属线程组: " + t1.getThreadGroup().getName());
System.out.println("主线程所属线程组: "
                + Thread.currentThread().getThreadGroup().getName());
```

程序运行结果如下，断点跟踪发现 t1 的线程组对象为 null：

```
TERMINATED
Exception in thread "main" java.lang.NullPointerException
at com.icss.aqs.TestGroup.main(TestGroup.java:77)
```

（4）调整线程创建方式，给线程 t1 增加要执行的任务。

```
ThreadGroup group = new ThreadGroup("worker");
Thread t1 = new Thread(group,new Runnable() {
    @Override
    public void run() {
        for(;;) {
            double x = Math.random()/Math.PI;
```

```
            }
        }
},"myThread");
t1.start();
try {
    Thread.sleep(500);
} catch (Exception e) {
}
System.out.println(t1.getState());
System.out.println("t1 线程所属线程组: " + t1.getThreadGroup().getName());
System.out.println("主线程所属线程组: "
            + Thread.currentThread().getThreadGroup().getName());
```

程序运行效果如下：

```
RUNNABLE
t1 线程所属线程组: worker
主线程所属线程组: main
```

反复运行上述的几个测试步骤，可以得到结论：如果线程 t1 的状态为 NEW 或 RUNNABLE，线程有所属线程组，如果线程状态为 TERMINATED，则它的所属线程组对象为 null。

（5）使用默认参数创建线程 t1（未指定线程组）。

```
Thread t1 = new Thread();
t1.start();
System.out.println("t1 线程所属线程组: " + t1.getThreadGroup().getName());
System.out.println("主线程所属线程组: "
            + Thread.currentThread().getThreadGroup().getName());
```

程序运行结果如下，t1 的线程组默认使用了创建者的线程组：

```
t1 线程所属线程组: main
主线程所属线程组: main
```

4.5.3 线程工厂接口

根据需要创建新的线程对象时，使用线程工厂可以避免 new Thread()的硬编码，使应用程序能够使用特殊的线程子类、线程优先级等，这样程序会更加灵活。

```
public interface ThreadFactory {
    Thread newThread(Runnable r);
}
```

线程工厂的简单实现参考如下：

```java
class SimpleThreadFactory implements ThreadFactory {
    public Thread newThread(Runnable r) {
        return new Thread(r);
    }
}
```

4.5.4 默认线程工厂实现

在 Executors 工具类中，提供了默认线程工厂的实现：

```java
public class Executors {
    public static ThreadFactory defaultThreadFactory() {
        return new DefaultThreadFactory();
    }
    static class DefaultThreadFactory implements ThreadFactory {
        private static final AtomicInteger poolNumber = new AtomicInteger(1);
        private final ThreadGroup group;
        private final AtomicInteger threadNumber = new AtomicInteger(1);
        private final String namePrefix;
        DefaultThreadFactory() {
            SecurityManager s = System.getSecurityManager();
            group = (s != null) ? s.getThreadGroup() :
                          Thread.currentThread().getThreadGroup();
            namePrefix = "pool-"+poolNumber.getAndIncrement()+"-thread-";
        }
        public Thread newThread(Runnable r) {
            Thread t = new Thread(group, r,
                    namePrefix + threadNumber.getAndIncrement(),0);
            if (t.isDaemon())
                t.setDaemon(false);
            if (t.getPriority() != Thread.NORM_PRIORITY)
                t.setPriority(Thread.NORM_PRIORITY);
            return t;
        }
    }
}
```

默认线程工厂的简单使用示例如下：

```java
public static void main(String[] args) {
    ThreadFactory factory = Executors.defaultThreadFactory();
    factory.newThread(new Runnable() {
```

```java
    public void run() {
        System.out.println("新线程: " + Thread.currentThread().getId());
    }
}).start();
System.out.println("主线程: " + Thread.currentThread().getId());
}
```

4.5.5 线程池与线程工厂

在创建线程池 ThreadPoolExecutor 的时候需要使用 ThreadFacotry 线程工厂。如果你没有指定一个线程工厂，那么 ThreadPoolExecutor 会使用默认的 ThreadFactory，即通过工具类 Executors 创建 defaultThreadFactory()对象。

```java
public class ThreadPoolExecutor
        extends AbstractExecutorService {
    private volatile ThreadFactory threadFactory;
    public ThreadPoolExecutor(int corePoolSize,int maximumPoolSize,
                        long keepAliveTime,TimeUnit unit,
                        BlockingQueue<Runnable> workQueue) {
        this(corePoolSize, maximumPoolSize, keepAliveTime, unit,
            workQueue,Executors.defaultThreadFactory(), defaultHandler);
    }
    public ThreadPoolExecutor(int corePoolSize,
                              int maximumPoolSize,
                              long keepAliveTime,
                              TimeUnit unit,
                              BlockingQueue<Runnable> workQueue,
                              ThreadFactory threadFactory,
                              RejectedExecutionHandler handler) {
        ...
        this.threadFactory = threadFactory;
    }
}
```

线程池 ThreadPoolExecutor 中的所有线程，都是在 addWorker()方法中，调用线程工厂 threadFactory 创建的。由于操作系统或用户策略对线程数量的限制，在创建新线程时可能会出现失败现象。创建线程时，需要分配本地栈的空间，应保证内存空间充分，特殊情况会出现内存溢出（OutOfMemory）错误。

4.6 线程池异常处理

线程池中的线程在执行任务时,也会出现异常情况,这时需要及时捕获异常并处理。如果未捕获异常,会导致当前线程终结,而且异常会抛出给最终用户。

对于未捕获的异常,可以采用 Thread.UncaughtExceptionHandler 接口统一处理,这样用户体验会更好。

4.6.1 异常捕获

线程池执行任务时,对于可能出现异常的地方,要及时使用 try/catch 捕获处理。

```java
public static void main(String[] args) {
    ExecutorService pool = Executors.newCachedThreadPool();
    for(int i=0;i<20;i++) {
        pool.execute(new Runnable() {
            public void run() {
                try {
                    ...    //及时捕获异常
                } catch (Exception e) {
                    ... //处理异常
                }
            }
        });
    }
    pool.shutdown();
}
```

4.6.2 UncaughtExceptionHandler 处理异常

Thread.UncaughtExceptionHandler 接口允许在每个 Thread 对象上都附着一个异常处理器。uncaughtException()方法会在线程因未捕获的异常而导致终结前调用。

```java
public class Thread {
    public interface UncaughtExceptionHandler {
        void uncaughtException(Thread t, Throwable e);
    }
}
```

可以调用 Thread 类的 setUncaughtExceptionHandler()方法为当前线程对象设置异常处理器。在 Thread 类中还可以使用 setDefaultUncaughtExceptionHandler()方法对所有线程对象设置默认异常处理器。

对线程池中的线程设置异常处理器，可以在构建线程池的时候传入一个自定义的 ThreadFactory 类，并且重写 ThreadFactory 类中 newThread(Runnable r)方法。

示例步骤如下：

（1）调用线程池，执行 20 个任务。当随机数大于 8 时，主动抛出异常。

```java
public static void main(String[] args) {
    ExecutorService pool = Executors.newCachedThreadPool();
    for(int i=0;i<20;i++) {
        pool.execute(new Runnable() {
            public void run() {
                int rand = (int)(Math.random()*10);
                if(rand > 8) {
                    throw new RuntimeException("test...");
                }
                System.out.println(Thread.currentThread().getId()
                                    + " running...");
            }
        });
    }
    pool.shutdown();
}
```

（2）由于异常未进行捕获，因此会抛出给调用者。程序运行效果如下（每次输出都会不同）。

```
11 running...
9 running...
14 running...
8 running...
15 running...
12 running...
8 running...
13 running...
13 running...
8 running...
Exception in thread "pool-1-thread-8" java.lang.RuntimeException: test...
    at java.lang.Thread.run(Thread.java:745)
Exception in thread "pool-1-thread-9" java.lang.RuntimeException: test...
    at java.lang.Thread.run(Thread.java:745)
12 running...
14 running...
17 running...
19 running...
```

```
10 running...
9 running...
11 running...
18 running...
```

（3）自定义异常捕获类，实现 Thread.UncaughtExceptionHandler 接口。使用 log4j 日志框架，记录捕获的异常信息。

```java
class MyUncaughtExceptionHandler
        implements Thread.UncaughtExceptionHandler {
    @Override
    public void uncaughtException(Thread t, Throwable e) {
        Log.logger.error(e.getMessage(),e);
    }
}
```

（4）自定义线程工厂，新创建的线程设置异常捕获方式。

```java
class MyThreadFactory implements ThreadFactory {
    @Override
    public Thread newThread(Runnable r) {
        Thread t = new Thread(r);
        t.setUncaughtExceptionHandler(new MyUncaughtExceptionHandler());
        return t;
    }
}
```

（5）修改主函数中线程池的创建方式，从 ThreadFactory 中创建新线程。

```java
public static void main(String[] args) {
    ExecutorService pool = Executors.newCachedThreadPool(new
                                            MyThreadFactory());
    for(int i=0;i<20;i++) {
        pool.execute(...);
    }
}
```

程序运行效果如下，所有线程池中未捕获的异常，Thread.UncaughtExceptionHandler 会捕获，然后记录在日志系统中。

```
13 running...
10 running...
8 running...
11 running...
25 running...
```

```
21 running...
17 running...
14 running...
28 running...
24 running...
20 running...
12 running...
15 running...
18 running...
19 running...
22 running...
23 running...
26 running...
ERROR - test...
java.lang.RuntimeException: test...
    at com.icss.pool.TaskExceptionHandle$1.run(TaskExceptionHandle.java:17)
    at java.lang.Thread.run(Thread.java:745)
ERROR - test...
java.lang.RuntimeException: test...
    at com.icss.pool.TaskExceptionHandle$1.run(TaskExceptionHandle.java:17)
    at java.lang.Thread.run(Thread.java:745)
```

总结：通过上述示例，使用 Thread.UncaughtExceptionHandler 和 ThreadFactory，可以有效避免线程池中的异常未经捕获处理，直接抛出给最终用户的现象，大幅提高了线程池的健壮性。

4.6.3　Future 处理异常

使用 Callable 与 Future 的组合，可以在 Future 返回结果中处理异常信息。Future 中存储了任务的运行结果，还可以取消任务以及检查任务是否发生了异常。Future 的规约中暗示了任务的生命周期是单向的，不能后退。就像 ExecutorService 的声明周期一样。一旦任务完成，它就永远停留在完成状态上。

任务的执行状态决定了 Future 的 get()方法行为。如果任务已经完成，get()方法会立即返回结果或者抛出一个异常；如果任务没有完成，get()方法会阻塞线程直到它完成。如果任务抛出了异常，get()方法会将该异常封装为 ExecutionException 异常，然后重新抛出。可以用 getCause()方法获得 ExecutionException 的异常信息。

示例操作步骤如下。

（1）自定义任务类，实现 Callable 接口，返回结果定义为 Integer 类型。

```java
class MyTaskCall implements Callable<Integer>{
    public Integer call() throws Exception {
        int rand = (int)(Math.random()*10);
```

```
        if(rand > 8) {
            throw new RuntimeException("test...");
        }
        System.out.println(Thread.currentThread().getId() + " running...");
        return rand;
    }
}
```

（2）使用线程池，执行异步任务。

```
public static void main(String[] args) {
    ExecutorService pool = Executors.newCachedThreadPool();
    for(int i=0;i<20;i++) {
        Future<Integer> future = pool.submit(new MyTaskCall());
        ...
    }
    pool.shutdown();
}
```

（3）调用 future.get()方法，返回任务执行结果。如果任务出现异常，用 ExecutionException 接收异常信息，并处理。

```
for(int i=0;i<20;i++) {
    Future<Integer> future = pool.submit(new MyTaskCall());
    try {
        Integer result = future.get();
        System.out.println(result);
    } catch (ExecutionException e) {
        Log.logger.error(e.getMessage(),e.getCause());
    }catch(Exception e) {
       System.out.println("网络异常，请检查...");
    }
}
```

程序运行结果如下，ExecutionException 捕获异常后，记录在日志文件中：

```
...
8 running...
6
8 running...
2
8 running...
0
8 running...
2
```

```
ERROR - java.lang.RuntimeException: test...
        at java.lang.Thread.run(Thread.java:745)
8 running...
5
8 running...
1
8 running...
3
```

4.7 本章习题

(1) 创建如下线程池对象,当并发任务数量为 10 时,会创建几个线程?(   )

```
BlockingQueue<Runnable> bq = new LinkedBlockingQueue<>(5);
ThreadPoolExecutor pool = new
        ThreadPoolExecutor(5,10,2000,TimeUnit.MILLISECONDS,bq);
```

A. 3 个线程　　　　B. 5 个线程　　　　C. 8 个线程　　　　D. 10 个线程

(2) 创建如下线程池对象,当并发任务数量为 13 时,会创建几个线程?(   )

```
BlockingQueue<Runnable> bq = new LinkedBlockingQueue<>(5);
ThreadPoolExecutor pool = new
        ThreadPoolExecutor(5,10,2000,TimeUnit.MILLISECONDS,bq);
```

A. 5 个线程　　　　B. 8 个线程　　　　C. 10 个线程　　　D. 13 个线程

(3) 线程池什么时候会抛出 RejectedExecutionException 异常?(   )

A. 当并发任务数大于 maximumPoolSize 时

B. 当并发任务数大于"corePoolSize +阻塞队列容量"时

C. 当任务数大于"maximumPoolSize+阻塞队列容量"时

D. 当并发任务数大于 corePoolSize 时

(4) 调用线程池 ThreadPoolExecutor 的什么方法可以执行异步任务?(   )

A. execute()方法　　B. run()方法　　　C. submit()方法　　D. invoke()方法

(5) 调用线程池 ThreadPoolExecutor 的什么方法可以关闭线程池?(   )

A. close()　　　　　B. shutdown()　　　C. stop()　　　　　D. shutdownNow()

(6) 调用线程池 ThreadPoolExecutor 的什么方法可以返回异步任务的执行结果?(   )

A. execute()方法　　B. run()方法　　　C. call()方法　　　D. submit()方法

(7) 当需要处理大量的短时任务时,应该使用 Executors 的哪个方法创建线程池?(   )

A. newFixedThreadPool()　　　　　　B. newScheduledThreadPool()
C. newCachedThreadPool()　　　　　 D. newWorkStealingPool()

（8）Web 服务器需要一个 socket 监听器，用下面哪种方法创建这个监听线程最佳？（　　）

A. new Thread(new Runnable(){})　　B. Executors.newSingleThreadExecutor()
C. 继承 Thread 类　　　　　　　　　D. 从线程组中获取

（9）Executors.newCachedThreadPool()创建的线程池，底层使用的阻塞队列是（　　）。

A. ArrayBlockingQueue　　　　　　　B. LinkedBlockingQueue
C. PriorityBlockingQueue　　　　　　D. SynchronousQueue

（10）Executors.newFixedThreadPool ()创建的线程池，底层使用的阻塞队列是（　　）。

A. ArrayBlockingQueue　　　　　　　B. LinkedBlockingQueue
C. PriorityBlockingQueue　　　　　　D. SynchronousQueue

（11）线程池 ThreadPoolExecutor 在执行任务时，如果出现异常，以下哪种方法处理得最好？（　　）

A. 把异常抛给主线程显示

B. 在 run()方法中，及时调用 try / catch 进行捕获处理

C. 使用 ExecutionException 捕获异常

D. 线程设置 UncaughtExceptionHandler 处理句柄，捕获异常

（12）创建一个新线程，如下哪个方式最佳？（　　）

A. new Thread(new Runnable(){})

B. 继承 Thread 类

C. 自定义线程工厂，然后调用 ThreadFactory. newThread()

D. 使用 Executors 的默认线程工厂，创建新线程

第 5 章 线程池与锁

本章将延续第 4 章的内容，对线程池中的锁机制做进一步的深入探索。

5.1 重入锁 ReentrantLock

ReentrantLock 翻译过来就是重入锁，读作 Re-entrant-Lock。

ReentrantLock 是一种线程互斥锁，它的基本行为和语义与使用 synchronized 的隐式监视器锁基本一致，但是更具有扩展性。所谓的互斥锁与重入锁概念，是指 ReentrantLock 对象在线程之间锁是互斥的，而同一个线程则可以反复调用 ReentrantLock 对象的 lock() 方法，进行多次加锁。

ReentrantLock 的推荐调用模式如下：

```java
class X {
    private final ReentrantLock lock = new ReentrantLock();
    public void m() {
        lock.lock();            //加锁
        try {
            // ... 方法体
        } finally {
            lock.unlock();      //解锁
        }
    }
}
```

这段代码需要注意以下两点：

（1）首先要创建一个 ReentrantLock 的成员对象实例，这是为了在该类的多个业务方法中调用同一个 ReentrantLock 对象的 lock() 方法。

（2）在每个业务方法的头部进行 lock() 锁定操作，在方法执行完成后，在 finally 中释放锁。由于 ReentrantLock 的锁必须手动释放（这与 synchronized 隐式锁不同），因此在 finally 中调用 unlock() 方法非常关键。一旦忘记锁的释放，其他线程想获得该锁时会出现长时间的等待。

5.1.1 重入锁

ReentrantLock 对象可以被同一线程反复递归调用,最大递归次数为 2 147 483 647。试图超出这个限制次数,将会导致从锁定方法中抛出 Error 异常。

ReentrantLock 重入调用的典型示例如下:

(1)定义类 LockX,在方法 m1()的方法体前加锁,在方法体尾部解锁。

```java
public class LockX {
    private final ReentrantLock lock = new ReentrantLock();
    public void m1() {
        lock.lock();                //加锁
        try {
            // ...方法体
        } finally {
            lock.unlock();          //解锁
        }
    }
}
```

(2)在类 LockX 中定义方法 m2(),方法体前后也分别使用 lock()与 unlock()进行加锁与解锁。

```java
public void m2() {
    lock.lock();                    //加锁
    try {
        // ... 方法体
    } finally {
        lock.unlock();              //解锁
    }
}
```

(3)在方法 m1()中调用方法 m2(),并跟踪当前线程对锁的持有数量。

```java
public void m1() {
    System.out.println("m1:" + lock.getHoldCount());
    lock.lock();                    //加锁
    try {
        // ...方法体
        System.out.println("m1:" + lock.getHoldCount());
        m2();                       //在m1()中调用m2()
    } finally {
        lock.unlock();              //解锁
```

```
        System.out.println("m1:" + lock.getHoldCount());
    }
}
public void m2() {
    lock.lock();                    //加锁
    try {
        // ... 方法体
        System.out.println("m2:" + lock.getHoldCount());
    } finally {
        lock.unlock();              //解锁
        System.out.println("m2:" + lock.getHoldCount());
    }
}
```

（4）在测试类中调用方法 m1()。

```
public static void main(String[] args) {
    LockX x = new LockX();
    x.m1();
}
```

程序运行结果如下，可以清楚地跟踪到锁持有者的数量。

```
m1:0
m1:1
m2:2
m2:1
m1:0
```

5.1.2 互斥锁

继续使用 5.1.1 节中的类 LockX，在日志中增加 Thread.currentThread().getId()，跟踪执行 m1()和 m2()方法的线程 ID。操作步骤如下：

（1）使用线程池，并发调用 LockX 对象的 m1()方法。

```
public static void main(String[] args) {
    LockX x = new LockX();
    ExecutorService pool = Executors.newFixedThreadPool(5);
    for(int i=0;i<5;i++) {
        pool.execute(new Runnable() {
            public void run() {
                x.m1();
            }
        });
```

```
    }
    pool.shutdown();
}
```

程序运行结果如下，5 个线程首先在 m1()方法的 lock.lock()处阻塞，争抢同一把锁。由于 ReentrantLock 为互斥锁，因此只能有一个线程进入 m1()的方法体。只有抢到锁的线程执行完毕，锁持有者数量为 0，第二个线程才有机会再次抢到锁。

```
m1:0,8
m1:0,9
m1:0,10
m1:0,11
m1:0,12
m1:1,8
m2:2,8
m2:1,8
m1:0,8
m1:1,9
m2:2,9
m2:1,9
m1:0,9
m1:1,10
m2:2,10
m2:1,10
m1:0,10
m1:1,11
m2:2,11
m2:1,11
m1:0,11
m1:1,12
m2:2,12
m2:1,12
m1:0,12
```

（2）把 m2()方法中的代码 lock.unlock()注释掉，程序运行结果如下，线程 8 首先抢到锁，然后进入 m1()和 m2()方法体，由于 m2()没有及时释放锁，因此所有的其他线程始终处于阻塞等待状态。

```
m1:0,11
m1:0,12
m1:0,8
m1:0,9
m1:0,10
```

```
m1:1,8
m2:2,8
m2:2,8
m1:1,8
```

重入锁还可以有效防止线程死锁。其简单机制是为每个锁关联一个请求计数器和一个占有它的线程对象，当计数器为 0 时，表示暂时没有线程占有它。当一个线程请求获得锁后，这个计数器会累加一次，当这个线程再次请求这个锁时，计数器再次累加一次。占有线程退出同步计数器递减一次，直到计数器为 0 时，这个锁才会被释放。

使用 ReentrantLock 进行加锁与解锁行为是显式的，我们加了多少次锁就要释放多少次锁。释放锁的行为一定要在 finally 中进行，这样可以保证锁一定会被释放。

5.1.3　ReentrantLock 与 synchronized

ReentrantLock 与 synchronized 在基本行为与语义上有很多相似性，下面对比分析一下两者的异同：

（1）它们都是互斥锁，都具有独占性和排他性。

（2）ReentrantLock 是重入锁，synchronized 也允许重入。

（3）ReentrantLock 是显式锁，需要显式地调用 lock() 与 unlock() 方法。synchronized 是隐式锁，加锁与解锁都依赖隐式的监视器。

（4）synchronized 是 Java 语言底层内置的锁机制，依赖于 JVM 直接解析字节码。而 ReentrantLock 是 JDK 层面的，可以查看其源代码和实现机制。

（5）ReentrantLock 应用更加灵活，具有很好的扩展性。调用 newCondition() 方法，可以创建 Condition 实例，这是用于多线程交互的条件监视器。另外为了防止由于忘记调用 unlock() 导致死锁，ReentrantLock 还提供了定时解锁功能。

synchronized 隐式锁重入与互斥特性测试如下：

（1）在类 LockY 中定义方法 m1() 和 m2()。在 m1() 与 m2() 中，用 synchronized 关键字锁同一个成员变量 object。在 m1() 方法中调用 m2() 方法。

```java
public class LockY {
    private Object object = new Object();
    public void m1() {
        synchronized(object) {
            //... 方法体
            System.out.println("m1:"+Thread.currentThread().getId()+"run...");
            m2();
        }
        System.out.println("m1:"+Thread.currentThread().getId()+"退出...");
    }
    public void m2() {
```

```java
    synchronized(object) {
        //... 方法体
        System.out.println("m2:"+ Thread.currentThread().getId()+"run...");
    }
    System.out.println("m2:" + Thread.currentThread().getId() + " 退出...");
    }
}
```

（2）在主函数中，创建 LockY 对象并调用 m1()。

```java
public static void main(String[] args) {
    LockY y = new LockY();
    y.m1();
}
```

程序运行结果如下，可以明确看出 synchronized 隐式锁的重入特性：

```
m1:1 run...
m2:1 run...
m2:1 退出...
m1:1 退出...
```

（3）并发调用测试。

```java
public static void main(String[] args) {
    LockY y = new LockY();
    ExecutorService pool = Executors.newFixedThreadPool(3);
    for(int i=0;i<3;i++) {
        pool.execute(new Runnable() {
            public void run() {
                y.m1();
            }
        });
    }
    pool.shutdown();
}
```

程序运行结果如下，可以明确看到线程互斥现象：

```
m1:8 run...
m2:8 run...
m2:8 退出...
m1:8 退出...
m1:10 run...
m2:10 run...
```

m2:10 退出...
m1:10 退出...
m1:9 run...
m2:9 run...
m2:9 退出...
m1:9 退出...

5.1.4 尝试加锁并限时等待

调用 tryLock()方法，可以尝试获取 ReentrantLock 的锁。如果当前线程已经保存该锁，则保持计数增加 1，该方法返回 true。如果锁由另一个线程持有，则该方法将立即返回 false。

tryLock(long timeout, TimeUnit unit)方法允许限时等待获取锁：如果在给定的等待时间内当前线程未被中断，且锁没有被其他线程持有，则获取该锁。

```java
public class ReentrantLock {
    public boolean tryLock() {
        return sync.nonfairTryAcquire(1);
    }
    public boolean tryLock(long timeout,TimeUnit unit)
                    throws InterruptedException{
        return sync.tryAcquireNanos(1, unit.toNanos(timeout));
    }
}
```

代码测试步骤如下：

（1）新建测试类，定义成员变量 ReentrantLock 对象。在方法 m1()中，调用 tryLock(5, TimeUnit.SECONDS)获取锁，超时则退出。

```java
public class TryLockTest {
    private ReentrantLock lock = new ReentrantLock();
    public void m1(){
        try {
            if (lock.tryLock(5, TimeUnit.SECONDS)) {
                System.out.println(Thread.currentThread().getId()+"获取了锁...");
                TimeUnit.SECONDS.sleep(3);
            } else {
                System.out.println(Thread.currentThread().getId()+"等待超时...");
            }
        }catch(Exception e){
```

```
            e.printStackTrace();
        }finally {
            if(lock.isHeldByCurrentThread() && lock.isLocked()) {
                lock.unlock();
            }
        }
    }
}
```

（2）并发调用 m1()方法，由于锁只能由一个线程获得，故其他线程需要等待。

```
public static void main(String[] args) {
    final TryLockTest test = new TryLockTest();
    ExecutorService pool = Executors.newFixedThreadPool(5);
    for(int i = 0; i < 5; i++) {
        pool.execute(new Runnable() {
            public void run() {
                test.m1();
            }
        });
    }
    pool.shutdown();
}
```

程序运行结果如下，超过等待时间的线程，会退出等待：

```
8 获取了锁...
9 获取了锁...
10 等待超时...
11 等待超时...
12 等待超时...
```

注意：调用 lock.unlock()方法时，需要进行判断；否则未获得锁的线程，调用 unlock()方法，会抛出 java.lang.IllegalMonitorStateException 异常。

```
if(lock.isHeldByCurrentThread() && lock.isLocked()) {
    lock.unlock();
}
```

尝试加锁并限时等待在某些场景下非常有用，它可以有效避免由于处理不当长时间占用资源，从而发生死锁的情况。如果某些线程由于异常等原因遗漏了 unlock()调用，也不会导致其他线程的无限期等待。

5.2 重入锁与 Condition

synchronized 关键字与 Object 对象的 wait()、notify()方法配合，可以实现多线程间的等待、唤醒操作，参见 1.3.4 节的代码示例。

ReentrantLock 锁与 Condition 条件对象配合，也可以实现多线程的等待/通知操作。

```java
public interface Condition {
    void await() throws InterruptedException;
    boolean await(long time,TimeUnit unit)
            throws InterruptedException;
    void signal();
    void signalAll();
}
```

Condition 对象的 await()方法，与 Object 对象的 wait()方法类似。而 Condition 对象的 signal()方法，与 Object 类中的 notify()方法类似。

把 1.3.4 节的代码进行修改，用 ReentrantLock 与 Condition 实现线程交互，操作步骤如下：

（1）主函数中创建 ReentrantLock 与 Condition 对象。

```java
public static void main(String[] args) {
    final ReentrantLock lock = new ReentrantLock();
    final Condition condition = lock.newCondition();
    ...
}
```

（2）定义线程对象 t1，当 i=5 时，阻塞当前线程。调用 condition.await()阻塞当前线程前，必须要调用 lock.lock()获取锁。

```java
Thread t1 = new Thread(new Runnable() {
    public void run() {
        for(int i=0;i<10;i++) {
            System.out.println(Thread.currentThread().getName()
                            + ",i=" +i);
            if(i==5) {
                lock.lock();
                System.out.println(Thread.currentThread().getName()
                            + "加锁");
                try {
                    System.out.println(Thread.currentThread()
                                .getName() + "开始等待...");
```

```java
                    condition.await();
                }catch(Exception e) {
                    e.printStackTrace();
                }finally {
                    lock.unlock();
                    System.out.println(Thread.currentThread()
                                    .getName() + "解锁");
                }
            }
        }
    }});
```

（3）定义线程对象 t2，把线程 t1 从 WAITING 状态唤醒。调用 condition.signal()唤醒 t1 线程前，必须要调用 lock.lock()获取锁。

```java
Thread t2 = new Thread(new Runnable() {
        public void run() {
            System.out.println(Thread.currentThread().getName()
                            + " running...");
            lock.lock();
            System.out.println(Thread.currentThread().getName()
                            + "加锁");
            try {
                System.out.println(Thread.currentThread().getName()
                                + ",发送notify通知...");
                condition.signal();
            } catch (Exception e) {
                e.printStackTrace();
            }finally {
                lock.unlock();
                System.out.println(Thread.currentThread().getName()
                                + "解锁");
            }
        }
}});
```

（4）分别启动线程 t1 和 t2。

```java
public static void main(String[] args) {
    final ReentrantLock lock = new ReentrantLock();
    final Condition condition = lock.newCondition();
    Thread t1 = new Thread(...);
    Thread t2 = new Thread(...);
    //启动第一个线程
    t1.start();
    double d = 0;
```

```
        for(int i=0;i<10000000;i++) {
                //等待第一个线程调用wait()方法
                d+=(Math.PI + Math.E)/(double)i;
        }
        //读取第一个线程的状态
        System.out.println(t1.getName()+ "状态: " + t1.getState());
        //启动第二个线程,发送notify()通知
        t2.start();
}
```

程序运行结果如下,这里需要特别注意的是在 t1 中调用 condition.await()后,t1 线程进入了 WAITING 状态,这时并没有调用 t1 中的 lock.unlock()。在 t1 并没有释放锁的情况下,t2 启动并调用 lock.lock()加锁成功。直到在 t2 中发出 condition.signal()通知,t1 才有机会解锁。

```
Thread-0,i=0
Thread-0,i=1
Thread-0,i=2
Thread-0,i=3
Thread-0,i=4
Thread-0,i=5
Thread-0 加锁
Thread-0 开始等待...
Thread-0 状态: WAITING
Thread-1 running...
Thread-1 加锁
Thread-1,发送 notify 通知...
Thread-1 解锁
Thread-0 解锁
Thread-0,i=6
Thread-0,i=7
Thread-0,i=8
Thread-0,i=9
```

总结:在调用 Object 对象的 wait()方法后,当前线程会释放 Object 对象的内置监视器,然后线程进入 WAITING 状态。这样,其他线程就可以通过 synchronized(object)重新获得对象监视器权。

同理,在调用 Condition 的 await()方法后,当前线程会释放 Condition 对象的监视器,然后线程进入 WAITING 状态。这样,其他线程调用 ReentrantLock 对象的 lock()方法,就会获取 ReentrantLock 对象的使用权。

synchronized 只能与一个对象的监视器关联,在某些场合下限制了程序的扩展性。而一个 ReentrantLock 对象,可以反复调用 newCondition()方法,创建多个 Condition 对象,这就

形成了一个条件队列。

5.2.1 案例分析：厨师与侍者

在 3.1.2 节，我们使用 synchronized 关键字与 Object 的 wait()和 notify()方法配合，实现了厨师与侍者的多线程交互。在本节，通过 ReentrantLock 结合 Condition 条件队列可以实现与 3.1.2 节同样的运行效果。

操作步骤如下：

（1）新建类 ChefWaiter，创建三个成员对象：一个 ReentrantLock 和两个 Condition 条件对象。

```java
public class ChefWaiter {
    private ReentrantLock lock = new ReentrantLock();
    private Condition chefCondition = lock.newCondition();
    private Condition waiterCondition = lock.newCondition();
}
```

（2）新建内部类 Chef，每次循环都要先使用重入锁进行 lock()，然后分别调用 chefCondition 和 waiterCondition 进行等待或通知操作。

```java
class Chef implements Runnable {
    private Waiter waiter;
    public void setWaiter(Waiter waiter) {
        this.waiter = waiter;
    }
    public void run() {
        while (true) {
            lock.lock();
            try {
                // chef 启动后，就处于等待 waiter 通知状态...
                chefCondition.await();
                System.out.println("厨师收到 waiter 下单后，开始做菜...");
                Thread.sleep(2000);
                System.out.println("厨师通知侍者取餐...");
                waiterCondition.signal();
            } catch (InterruptedException e) {
                e.printStackTrace();
            } finally {
                lock.unlock();
            }
        }
    }
}
```

（3）新建内部类 Waiter,每次循环，都要先使用重入锁进行 lock()，然后分别调用 chefCondition 和 waiterCondition 进行等待或通知操作。

```java
class Waiter implements Runnable {
    private Chef chef;
    public void setChef(Chef chef) {
        this.chef = chef;
    }
    public void run() {
        while (true) {
            lock.lock();
            try {
                Thread.sleep(1000);
                System.out.println("waiter 接到下单,通知厨师做菜");
                chefCondition.signal();
                // 通知厨师后，waiter 进入等待顾客下单状态
                waiterCondition.await();
                System.out.println("接到厨师通知，侍者取餐");
            } catch (Exception e) {
                e.printStackTrace();
            } finally {
                lock.unlock();
            }
        }
    }
}
```

（4）新建 ChefWaiter 的成员方法，分别启动两个线程运行 Chef 与 Waiter。

```java
public void startWork() {
    Chef chef = new Chef();
    Waiter waiter = new Waiter();
    chef.setWaiter(waiter);
    waiter.setChef(chef);
    new Thread(chef).start();
    new Thread(waiter).start();
}
```

（5）在主函数中启动程序。

```java
public static void main(String[] args) {
    ChefWaiter cw = new ChefWaiter();
    cw.startWork();
}
```

程序运行结果，与 3.1.2 节完全一致：

```
waiter 接到下单,通知厨师做菜
厨师收到 waiter 下单后，开始做菜...
厨师通知侍者取餐...
接到厨师通知，侍者取餐
waiter 接到下单,通知厨师做菜
厨师收到 waiter 下单后，开始做菜...
厨师通知侍者取餐...
接到厨师通知，侍者取餐
waiter 接到下单,通知厨师做菜
厨师收到 waiter 下单后，开始做菜...
厨师通知侍者取餐...
接到厨师通知，侍者取餐
waiter 接到下单,通知厨师做菜
厨师收到 waiter 下单后，开始做菜...
厨师通知侍者取餐...
接到厨师通知，侍者取餐
```

5.2.2 案例分析：缓冲区队列

假设需要一个空间有限的缓冲区队列，它支持 put()方法写入数据和 take()方法取出数据。如果在一个空的缓冲区尝试提取元素，则线程将阻塞等待，直到队列有可用的元素可以提取。如果试图在一个已满的缓冲区写入数据，那么线程也会阻塞，直到队列空间变得可用。这个缓冲区队列支持多线程并发读写，为了提高并发性能，可以使用两个 Condition 对象来控制。

操作步骤如下：

（1）新建缓冲类 BoundedBuffer。同时创建一个重入锁对象和两个 Condition 对象。Object[] 数组用于存储缓冲数据。

```java
public class BoundedBuffer {
    final Lock lock = new ReentrantLock();
    final Condition notFull = lock.newCondition();
    final Condition notEmpty = lock.newCondition();
    final Object[] items = new Object[20];
    int putptr, takeptr, count;
}
```

（2）新建 put()方法用于向队列中写入数据。当队列已满时，调用 Condition 的 await()方法阻塞写入数据的线程。

```java
public void put(Object x) throws InterruptedException {
```

```java
        lock.lock();
        try {
            while (count == items.length) {
                System.out.println(Thread.currentThread().getId() + ",队列满了...");
                notFull.await();
            }
            items[putptr] = x;
            if (++putptr == items.length)
                putptr = 0;
            ++count;
            notEmpty.signal();
        } finally {
            lock.unlock();
        }
    }
```

（3）新建 take()方法用于从队列中取出数据。当队列为空时，调用另外一个 Condition 对象的 await()方法，阻塞读取数据的线程。

```java
    public Object take() throws InterruptedException {
        lock.lock();
        try {
            while (count == 0) {
                System.out.println(Thread.currentThread().getId() + ",队列空了...");
                notEmpty.await();
            }
            Object x = items[takeptr];
            if (++takeptr == items.length)
                takeptr = 0;
            --count;
            notFull.signal();
            return x;
        } finally {
            lock.unlock();
        }
    }
```

（4）在主函数中，分别创建一个写数据的线程和一个读数据的线程，两个线程同时工作。将写入数据的延迟时间调得稍短一些，则很容易出现队列已满、线程阻塞等待的现象。将取出数据的延迟时间调得稍短一些，则很容易出现队列为空，线程阻塞等待的现象。

```java
    public static void main(String[] args){
        BoundedBuffer buffer = new BoundedBuffer();
        new Thread(new Runnable() {
```

```java
        public void run() {
            while(true) {
                int rand = (int)(Math.random()*100);
                try {
                    buffer.put(rand);
                    System.out.println(Thread.currentThread().getId()
                                    + "写入: " + rand);
                    Thread.sleep(100);
                } catch (Exception e) {
                    e.printStackTrace();
                }
            }
        }
    }).start();
    new Thread(new Runnable() {
        public void run() {
            while(true) {
                try {
                    int ii = (int)buffer.take();
                    System.out.println(Thread.currentThread().getId()
                                    + "取出: " + ii);
                    Thread.sleep(100);
                } catch (Exception e) {
                    e.printStackTrace();
                }
            }
        }
    }).start();
}
```

程序运行结果如下：

8 写入: 11
8 写入: 74
9 取出: 45
8 写入: 64
8 写入: 32
9 取出: 58
8 写入: 42
8 写入: 76
9 取出: 7
8 写入: 47
8,队列满了...
9 取出: 7

```
8 写入：6
8,队列满了...
9 取出：41
8 写入：63
8,队列满了...
8 写入：12
9 取出：37
8,队列满了...
8 写入：33
9 取出：33
9,队列空了...
8 写入：47
9 取出：47
9,队列空了...
8 写入：15
9 取出：15
9,队列空了...
8 写入：82
9 取出：82
9,队列空了...
```

5.3 读锁与写锁

synchronized 隐式锁和 ReentrantLock 重入锁都具有互斥性，即同一时间只能有一个线程获得锁，其他线程若想取得相同对象的锁，需要排队等候。

在使用 ReentrantReadWriteLock 读写锁时，一个写锁可以降级为一个读锁，但是读锁不能成为一个写锁；写操作时，其他线程是可读的，但是读取时，其他线程不能写入。

参阅 ReentrantReadWriteLock 的源代码，其内部设置了两把锁：一个是 ReadLock 读锁对象，一个 WriteLock 写锁对象。ReadLock 读锁中使用的是 acquireShared()方法，表示可用共享的模式获取锁。WriteLock 写锁中使用的是 acquire()方法，表示以独占模式获取锁。acquireShared()方法和 acquire()方法都属于 AbstractQueuedSynchronizer 类中的方法，关于 AbstractQueuedSynchronizer 类将会在后面章节中介绍。

ReentrantReadWriteLock 源码片段如下所示：

```java
public class ReentrantReadWriteLock
        implements ReadWriteLock, java.io.Serializable {
    private final ReentrantReadWriteLock.ReadLock readerLock;
    private final ReentrantReadWriteLock.WriteLock writerLock;
    final Sync sync;
    ...
```

}
```

ReentrantReadWriteLock 有两个静态内部类，分别为 ReadLock 和 WriteLock。

```java
public static class ReadLock
        implements Lock, java.io.Serializable {
    private final Sync sync;
    protected ReadLock(ReentrantReadWriteLock lock) {
        sync = lock.sync;
    }
    public void lock() {
        sync.acquireShared(1);
    }
    public void unlock() {
        sync.releaseShared(1);
    }
    public Condition newCondition() {
        throw new UnsupportedOperationException();
    }
    ...
}
public static class WriteLock
        implements Lock, java.io.Serializable {
    private final Sync sync;
    protected WriteLock(ReentrantReadWriteLock lock) {
        sync = lock.sync;
    }
    public void lock() {
        sync.acquire(1);
    }
    public void unlock() {
        sync.release(1);
    }
    public Condition newCondition() {
        return sync.newCondition();
    }
    ...
}
```

5.3.1 案例：并发读写集合

集合 ArrayList 在并发读写时，会出现各种异常，参见如下代码测试。
（1）使用两个线程池，分别对同一个 ArrayList 集合进行读和写的操作。
```

```java
public static void main(String[] args) {
    List<Book> bookList = new ArrayList<Book>();
    ExecutorService readPool = Executors.newFixedThreadPool(2);
    ExecutorService writePool = Executors.newFixedThreadPool(2);
    for(int i=0;i<2;i++) {
        writePool.execute(new Runnable() {
            public void run() {
                for(int i=0;i<20;i++) {
                    Book bk = new Book(Thread.currentThread().getId()
                                + "写入" + i);
                    bookList.add(bk);
                    System.out.println(bk.getBname());
                }
            }
        });
    }
    for(int i=0;i<2;i++) {
        readPool.execute(new Runnable() {
            public void run() {
                for(Book bk : bookList) {
                    System.out.println(Thread.currentThread().getId()
                                + "读取: " + bk.getBname());
                }
            }
        });
    }
    readPool.shutdown();
    writePool.shutdown();
}
```

程序运行结果如下，很容易出现 ConcurrentModificationException 异常。

8 写入 0
8 写入 1
8 写入 2
8 写入 3
8 写入 4
8 写入 5
8 写入 6
8 写入 7
8 写入 8
8 写入 9
8 写入 10

```
8 写入 11
8 写入 12
8 写入 13
8 写入 14
8 写入 15
8 写入 16
8 写入 17
8 写入 18
8 写入 19
10 读取：8 写入 0
9 写入 0
9 写入 1
9 写入 2
9 写入 3
9 写入 4
9 写入 5
9 写入 6
9 写入 7
9 写入 8
9 写入 9
9 写入 10
9 写入 11
9 写入 12
9 写入 13
9 写入 14
9 写入 15
9 写入 16
9 写入 17
9 写入 18
9 写入 19
Exception in thread "pool-1" java.util.ConcurrentModificationException
    at java.util.ArrayList$Itr.checkForComodification(ArrayList.java:901)
    at java.util.ArrayList$Itr.next(ArrayList.java:851)
```

（2）创建 ReadWriteLock 对象。

```java
public static void main(String[] args) {
    ReadWriteLock lock = new ReentrantReadWriteLock();
    ...
}
```

（3）向集合中写入数据时，用写锁 ReentrantReadWriteLock.WriteLock 进行控制。

```java
for(int i=0;i<2;i++) {
```

```java
    writePool.execute(new Runnable() {
        public void run() {
            lock.writeLock().lock();
            for(int i=0;i<20;i++) {
                Book bk = new Book(Thread.currentThread().getId()
                                + "写入" + i);
                bookList.add(bk);
                System.out.println(bk.getBname());
            }
            lock.writeLock().unlock();
        }
    });
}
```

(4) 从集合中读取数据时,用 ReentrantReadWriteLock.ReadLock 进行读锁控制。

```java
for(int i=0;i<2;i++) {
    readPool.execute(new Runnable() {
        public void run() {
            lock.readLock().lock();
            for(Book bk : bookList) {
                System.out.println(Thread.currentThread().getId()
                                + "读取: " + bk.getBname());
            }
            lock.readLock().unlock();
        }
    });
}
```

程序运行结果如下,由于集合的读与写都使用 ReentrantReadWriteLock 进行了读写分离的控制,故不会再出现异常。注意:由于是线程抢占模式,用于读的线程 11 可能会抢在用于写的线程 9 之前执行,也可能在其后执行。

```
8 写入 0
8 写入 1
8 写入 2
8 写入 3
8 写入 4
8 写入 5
8 写入 6
8 写入 7
8 写入 8
8 写入 9
8 写入 10
```

8 写入 11
8 写入 12
8 写入 13
8 写入 14
8 写入 15
8 写入 16
8 写入 17
8 写入 18
8 写入 19
11 读取：8 写入 0
11 读取：8 写入 1
11 读取：8 写入 2
11 读取：8 写入 3
11 读取：8 写入 4
11 读取：8 写入 5
11 读取：8 写入 6
11 读取：8 写入 7
11 读取：8 写入 8
11 读取：8 写入 9
11 读取：8 写入 10
11 读取：8 写入 11
11 读取：8 写入 12
11 读取：8 写入 13
11 读取：8 写入 14
11 读取：8 写入 15
11 读取：8 写入 16
11 读取：8 写入 17
11 读取：8 写入 18
11 读取：8 写入 19
9 写入 0
9 写入 1
9 写入 2
9 写入 3
9 写入 4
9 写入 5
9 写入 6
9 写入 7
9 写入 8
9 写入 9
9 写入 10
9 写入 11
9 写入 12
9 写入 13

9 写入 14
9 写入 15
9 写入 16
9 写入 17
9 写入 18
9 写入 19
10 读取：8 写入 0
10 读取：8 写入 1
10 读取：8 写入 2
10 读取：8 写入 3
10 读取：8 写入 4
10 读取：8 写入 5
10 读取：8 写入 6
10 读取：8 写入 7
10 读取：8 写入 8
10 读取：8 写入 9
10 读取：8 写入 10
10 读取：8 写入 11
10 读取：8 写入 12
10 读取：8 写入 13
10 读取：8 写入 14
10 读取：8 写入 15
10 读取：8 写入 16
10 读取：8 写入 17
10 读取：8 写入 18
10 读取：8 写入 19
10 读取：9 写入 0
10 读取：9 写入 1
10 读取：9 写入 2
10 读取：9 写入 3
10 读取：9 写入 4
10 读取：9 写入 5
10 读取：9 写入 6
10 读取：9 写入 7
10 读取：9 写入 8
10 读取：9 写入 9
10 读取：9 写入 10
10 读取：9 写入 11
10 读取：9 写入 12
10 读取：9 写入 13
10 读取：9 写入 14
10 读取：9 写入 15
10 读取：9 写入 16

```
10 读取：9 写入 17
10 读取：9 写入 18
10 读取：9 写入 19
```

（5）在程序的读写之间增加延迟，即保证集合并发写之后，再进行并发读操作。

```
for(int i=0;i<2;i++) {...写...}
try {
    Thread.sleep(1000);
} catch (Exception e) {
}
for(int i=0;i<2;i++) {...读...}
```

程序运行结果如下，从运行结果可以非常明确地验证前面的结论，即写锁 ReentrantReadWriteLock.WriteLock 是排他的，所以线程 8 和线程 9 在并发写入时，必须是顺序执行。而读锁 ReentrantReadWriteLock.ReadLock 是可以共享的，所以线程 10 和线程 11 可以交替运行输出。

```
9 写入 0
9 写入 1
9 写入 2
9 写入 3
9 写入 4
9 写入 5
9 写入 6
9 写入 7
9 写入 8
9 写入 9
9 写入 10
9 写入 11
9 写入 12
9 写入 13
9 写入 14
9 写入 15
9 写入 16
9 写入 17
9 写入 18
9 写入 19
8 写入 0
8 写入 1
8 写入 2
8 写入 3
8 写入 4
8 写入 5
```

8 写入 6
8 写入 7
8 写入 8
8 写入 9
8 写入 10
8 写入 11
8 写入 12
8 写入 13
8 写入 14
8 写入 15
8 写入 16
8 写入 17
8 写入 18
8 写入 19
10 读取：9 写入 0
10 读取：9 写入 1
10 读取：9 写入 2
10 读取：9 写入 3
10 读取：9 写入 4
10 读取：9 写入 5
10 读取：9 写入 6
10 读取：9 写入 7
10 读取：9 写入 8
10 读取：9 写入 9
10 读取：9 写入 10
10 读取：9 写入 11
10 读取：9 写入 12
10 读取：9 写入 13
10 读取：9 写入 14
10 读取：9 写入 15
10 读取：9 写入 16
10 读取：9 写入 17
10 读取：9 写入 18
10 读取：9 写入 19
11 读取：9 写入 0
11 读取：9 写入 1
11 读取：9 写入 2
11 读取：9 写入 3
10 读取：8 写入 0
11 读取：9 写入 4
10 读取：8 写入 1
11 读取：9 写入 5
10 读取：8 写入 2

```
11 读取：9 写入 6
10 读取：8 写入 3
11 读取：9 写入 7
10 读取：8 写入 4
11 读取：9 写入 8
10 读取：8 写入 5
11 读取：9 写入 9
10 读取：8 写入 6
11 读取：9 写入 10
10 读取：8 写入 7
11 读取：9 写入 11
10 读取：8 写入 8
11 读取：9 写入 12
10 读取：8 写入 9
11 读取：9 写入 13
10 读取：8 写入 10
11 读取：9 写入 14
10 读取：8 写入 11
11 读取：9 写入 15
10 读取：8 写入 12
11 读取：9 写入 16
10 读取：8 写入 13
11 读取：9 写入 17
10 读取：8 写入 14
11 读取：9 写入 18
10 读取：8 写入 15
11 读取：9 写入 19
10 读取：8 写入 16
11 读取：8 写入 0
10 读取：8 写入 17
11 读取：8 写入 1
10 读取：8 写入 18
11 读取：8 写入 2
10 读取：8 写入 19
11 读取：8 写入 3
11 读取：8 写入 4
11 读取：8 写入 5
11 读取：8 写入 6
11 读取：8 写入 7
11 读取：8 写入 8
11 读取：8 写入 9
11 读取：8 写入 10
11 读取：8 写入 11
```

11 读取：8 写入 12
11 读取：8 写入 13
11 读取：8 写入 14
11 读取：8 写入 15
11 读取：8 写入 16
11 读取：8 写入 17
11 读取：8 写入 18
11 读取：8 写入 19

（6）删除步骤（5）的延迟，在每次读写之间增加延迟，即确保读写线程同时运行。

```java
public void run() {
    lock.writeLock().lock();
    for(int i=0;i<20;i++) {
        Book bk = new Book(Thread.currentThread().getId() + "写入" + i);
        bookList.add(bk);
        System.out.println(bk.getBname());
        try {
            Thread.sleep(100);
        } catch (Exception e) {    }
    }
    lock.writeLock().unlock();
}
public void run() {
    lock.readLock().lock();
    for(Book bk : bookList) {
        System.out.println(Thread.currentThread().getId()
                        + "读取：" + bk.getBname());
        try {
            Thread.sleep(150);
        } catch (Exception e) {    }
    }
    lock.readLock().unlock();
}
```

程序运行结果如下（每次运行的效果会不同），反复运行步骤（6），可以得到如下结论：读锁与写锁之间是互斥的，读锁与读锁之间可以并行。

8 写入 0
8 写入 1
8 写入 2
8 写入 3
8 写入 4
8 写入 5

8 写入 6
8 写入 7
8 写入 8
8 写入 9
8 写入 10
8 写入 11
8 写入 12
8 写入 13
8 写入 14
8 写入 15
8 写入 16
8 写入 17
8 写入 18
8 写入 19
10 读取：8 写入 0
11 读取：8 写入 0
10 读取：8 写入 1
11 读取：8 写入 1
10 读取：8 写入 2
11 读取：8 写入 2
10 读取：8 写入 3
11 读取：8 写入 3
10 读取：8 写入 4
11 读取：8 写入 4
10 读取：8 写入 5
11 读取：8 写入 5
10 读取：8 写入 6
11 读取：8 写入 6
10 读取：8 写入 7
11 读取：8 写入 7
10 读取：8 写入 8
11 读取：8 写入 8
10 读取：8 写入 9
11 读取：8 写入 9
10 读取：8 写入 10
11 读取：8 写入 10
10 读取：8 写入 11
11 读取：8 写入 11
10 读取：8 写入 12
11 读取：8 写入 12
10 读取：8 写入 13
11 读取：8 写入 13
10 读取：8 写入 14

11 读取：8 写入 14
10 读取：8 写入 15
11 读取：8 写入 15
10 读取：8 写入 16
11 读取：8 写入 16
10 读取：8 写入 17
11 读取：8 写入 17
10 读取：8 写入 18
11 读取：8 写入 18
10 读取：8 写入 19
11 读取：8 写入 19
9 写入 0
9 写入 1
9 写入 2
9 写入 3
9 写入 4
9 写入 5
9 写入 6
9 写入 7
9 写入 8
9 写入 9
9 写入 10
9 写入 11
9 写入 12
9 写入 13
9 写入 14
9 写入 15
9 写入 16
9 写入 17
9 写入 18
9 写入 19

5.3.2 案例：Map 并发控制

JDK 中的 HashMap 与 TreeMap 都没有并发控制，因此在 ReentrantReadWriteLock 的 API 帮助中，为我们提供了一个用锁封装 Map 集合的办法。

操作步骤如下：

（1）定义类 RWHashMap，继承的父类与接口都与 HashMap 相同。

```
public class RWHashMap<K,V> extends AbstractMap<K,V>
            implements Map<K,V>, Cloneable, Serializable {
}
```

（2）采用聚合模式创建成员变量 hashMap，用这个对象实现业务方法的操作。另外创建了 ReentrantReadWriteLock 对象，并分别提取了读锁和写锁。

```java
private HashMap<K,V> hashMap;
private final ReentrantReadWriteLock rw
                    = new ReentrantReadWriteLock();
private final Lock r = rw.readLock();
private final Lock w = rw.writeLock();
```

（3）重新包装 HashMap 的构造函数。

```java
public RWHashMap(int initialCapacity) {
    hashMap = new HashMap<K,V>(initialCapacity);
}
```

（4）使用读锁，包装 get()方法和 entrySet()方法。

```java
public V get(Object key) {
    r.lock();
    try {
        return hashMap.get(key);
    } finally {
        r.unlock();
    }
}
public Set<Entry<K, V>> entrySet() {
    r.lock();
    try {
        return hashMap.entrySet();
    } finally {
        r.unlock();
    }
}
```

（5）使用写锁，包装 put()方法。

```java
public V put(K key, V value) {
    w.lock();
    try {
        return hashMap.put(key, value);
    } finally {
        w.unlock();
    }
}
```

（6）在主函数中，采用并发模式，对 RWHashMap 进行测试。

```java
public static void main(String[] args) {
    ExecutorService readPool = Executors.newFixedThreadPool(2);
    ExecutorService writePool = Executors.newFixedThreadPool(2);
    RWHashMap<String,Integer> rwHashMap = new RWHashMap<String,Integer>(16);
    for(int i=0;i<2;i++) {
        writePool.execute(new Runnable() {
            public void run() {
                for(int i=0;i<20;i++) {
                    String key = Thread.currentThread().getId() + "-" + i;
                    rwHashMap.put(key, i);
                    System.out.println(Thread.currentThread().getId()
                            + "写入：" + i);
                }
            }
        });
    }
    for(int i=0;i<2;i++) {
        readPool.execute(new Runnable() {
            public void run() {
                for(Map.Entry<String, Integer> entry : rwHashMap.entrySet()){
                    System.out.println(Thread.currentThread().getId()
                            + "读：" + entry.getKey());
                }
            }
        });
    }
    readPool.shutdown();
    writePool.shutdown();
}
```

测试结果如下，这与我们预期的并发效果差别很大。Map 集合的写与读之间出现了交错，甚至出现了 ConcurrentModificationException 异常。实际上，采用本节示例中的在 Map 集合中内置 ReentrantReadWriteLock 的方法，只能解决单个方法的读与写的并发问题。当采用循环方式写入时，每调用一次 put()方法，就需要重新获取一次写锁，这就给了读锁插队的机会。只有采用 5.3.1 节中的示例方式，在循环外面加锁，才能从根本上解决并发冲突问题。

9 写入：0
9 写入：1
9 写入：2
9 写入：3

```
9 写入：4
8 写入：0
8 写入：1
8 写入：2
8 写入：3
8 写入：4
8 写入：5
10 读：9-4
10 读：8-5
10 读：9-5
8 写入：6
8 写入：7
8 写入：8
8 写入：9
8 写入：10
8 写入：11
8 写入：12
8 写入：13
8 写入：14
8 写入：15
8 写入：16
8 写入：17
8 写入：18
8 写入：19
11 读：8-16
11 读：8-17
11 读：8-14
11 读：8-15
11 读：8-18
11 读：8-19
11 读：8-12
11 读：8-13
11 读：8-10
11 读：8-11
11 读：8-0
11 读：9-0
11 读：8-1
11 读：9-1
9 写入：5
9 写入：6
11 读：8-2
Exception in thread "pool-1-thread-1" ConcurrentModificationException at
java.util.HashMap$HashIterator.nextNode(HashMap.java:1429)
```

```
at java.lang.Thread.run(Thread.java:745)
9 写入: 7
9 写入: 8
9 写入: 9
9 写入: 10
9 写入: 11
9 写入: 12
9 写入: 13
9 写入: 14
9 写入: 15
9 写入: 16
9 写入: 17
9 写入: 18
9 写入: 19
```

5.3.3 数据库事务与锁

读锁又被称为共享锁，简称 S 锁，是解决数据库的脏读、不可重复读问题的重要手段。写锁又被称为排他锁，简称 X 锁。写锁与写锁之间互斥，写锁与读锁之间也是互斥的，即正在读锁保护下读取数据时，是不允许修改数据的。读锁与读锁之间是共享的，即事务 A 在读取共享数据时，事务 B 也可以读取该数据。

在关系型数据库中，设置了 4 个事务隔离级别，它们分别为 Read uncommitted、Read committed、Repeatable Read 和 Serializable。Oracle 数据库默认为 read committed 隔离，MySQL 数据库默认为 repeatable read 隔离。

- Read uncommitted：性能最佳，安全性最差，有脏读。
- Read committed：解决了脏读，会出现不可重复读，幻读。
- Repeatable Read：解决了重复读，会出现幻读。
- Serializable：解决了幻读，性能最差，但安全性最高。

关系型数据库的事务隔离级别，是依赖 S 锁和 X 锁实现的，具体加锁情况如下。

Read uncommitted，使用一级锁协议：

事务在读数据时并未对数据加锁。

事务在修改数据时只对数据增加使用行级共享锁，事务结束后释放。

Read committed，使用二级锁协议：

事务在读数据时，先加行级共享锁，读完就释放。

事务在修改数据时，先加行级排他锁，事务结束后释放。

Reapeatable Read，使用三级锁协议：

事务在读数据时，先加行级共享锁，事务结束后释放。

事务在修改数据时，先加行级排他锁，事务结束后释放。

Serializable，使用四级锁协议：

事务在读数据时，先加表级共享锁，事务结束后释放。

事务在修改数据时，先加表级排他锁，事务结束后释放。

5.4 公平锁与非公平锁

ReentrantLock 对象创建时默认使用非公平锁模式，在其构造函数中输入参数 true，则会使用公平锁模式进行任务排队。

在非公平锁模式下，新的任务线程与已经排队的任务线程竞争，可能会优先获得 ReentrantLock 锁的使用权。而在公平锁模式下，所有任务线程必须老老实实地排队，先提交的任务优先获得 ReentrantLock 锁的使用权。

非公平锁在高并发下，性能更优。因此，除非对任务的执行顺序有特殊要求，否则推荐使用默认的非公平锁模式。

公平锁与非公平锁代码测试步骤如下：

（1）定义任务类，所有任务执行时共用同一个 ReentrantLock 对象锁。

```java
class MyTask implements Runnable{
    private String name;
    private ReentrantLock locker;
    public MyTask(String name,ReentrantLock locker) {
        this.name = name;
        this.locker = locker;
    }
    public void run() {
        locker.lock();
        try {
            Log.logger.info(this.name + "--运行在线程"
                    + Thread.currentThread().getId());
        } finally {
            locker.unlock();
        }
    }
}
```

（2）在主函数，使用默认的非公平锁执行任务。

```java
public static void main(String[] args) {
    ReentrantLock locker = new ReentrantLock(false);
    ExecutorService pool = Executors.newFixedThreadPool(3);
    for(int i=0;i<30;i++) {
        try {
```

```
                pool.execute(new MyTask(Integer.toString(i),locker));
                Thread.sleep(5);
            } catch (Exception e) {
            }
        }
        pool.shutdown();
    }
```

程序运行效果如下，输出效果与任务提交的顺序并不一致，说明非公平锁模式后面提交的任务抢占了前面任务的位置。

```
INFO - 0--运行在线程 8
INFO - 1--运行在线程 9
INFO - 4--运行在线程 9
INFO - 2--运行在线程 10
INFO - 6--运行在线程 10
INFO - 7--运行在线程 10
INFO - 8--运行在线程 10
INFO - 9--运行在线程 10
INFO - 3--运行在线程 8
INFO - 11--运行在线程 8
INFO - 5--运行在线程 9
INFO - 13--运行在线程 9
INFO - 14--运行在线程 9
INFO - 15--运行在线程 9
INFO - 16--运行在线程 9
INFO - 17--运行在线程 9
INFO - 18--运行在线程 9
INFO - 19--运行在线程 9
INFO - 20--运行在线程 9
INFO - 21--运行在线程 9
INFO - 22--运行在线程 9
INFO - 23--运行在线程 9
INFO - 24--运行在线程 9
INFO - 25--运行在线程 9
INFO - 26--运行在线程 9
INFO - 27--运行在线程 9
INFO - 10--运行在线程 10
INFO - 29--运行在线程 10
INFO - 12--运行在线程 8
INFO - 28--运行在线程 9
```

（3）在主函数，使用公平锁执行任务。

```java
public static void main(String[] args) {
    ReentrantLock locker = new ReentrantLock(true);
    ExecutorService pool = Executors.newFixedThreadPool(3);
    for(int i=0;i<30;i++) {
        try {
                pool.execute(new MyTask(Integer.toString(i),locker));
                Thread.sleep(5);
        } catch (Exception e) {
        }
    }
    pool.shutdown();
}
```

程序运行效果如下,输出内容与任务的提交顺序完全一致。

```
INFO  -  0--运行在线程 8
INFO  -  1--运行在线程 9
INFO  -  2--运行在线程 10
INFO  -  3--运行在线程 8
INFO  -  4--运行在线程 9
INFO  -  5--运行在线程 10
INFO  -  6--运行在线程 8
INFO  -  7--运行在线程 9
INFO  -  8--运行在线程 10
INFO  -  9--运行在线程 8
INFO  -  10--运行在线程 9
INFO  -  11--运行在线程 10
INFO  -  12--运行在线程 8
INFO  -  13--运行在线程 9
INFO  -  14--运行在线程 10
INFO  -  15--运行在线程 8
INFO  -  16--运行在线程 9
INFO  -  17--运行在线程 10
INFO  -  18--运行在线程 8
INFO  -  19--运行在线程 9
INFO  -  20--运行在线程 10
INFO  -  21--运行在线程 8
INFO  -  22--运行在线程 9
INFO  -  23--运行在线程 10
INFO  -  24--运行在线程 8
INFO  -  25--运行在线程 9
INFO  -  26--运行在线程 10
INFO  -  27--运行在线程 8
INFO  -  28--运行在线程 9
```

```
INFO - 29--运行在线程 10
```

注意：上述公平锁与非公平锁的测试中，任务执行时间设定的是 5 毫秒。如果任务执行时间设置得过长，则非公平锁的抢占效果不明显。

5.5 本章习题

（1）ReentrantLock 与 synchronized 关键字在使用中的区别，如下描述正确的是（　　）。
A. ReentrantLock 锁可以反复重入，synchronized 锁对象不可以重入
B. ReentrantLock 是显式锁，需要显式地加锁与释放；synchronized 是隐式锁，加锁与解锁，都依赖隐式的监视器
C. synchronized 锁具有独占性和排他性，ReentrantLock 可以创建共享锁
D. synchronized 锁比 ReentrantLock 锁更加灵活，扩展性更好

（2）如下关于 Condition 的描述正确的是（　　）。
A. 一个 ReentrantLock 对象，只能创建一个 Condition 对象
B. 调用 Condition 对象的 await()方法或 Object 对象的 wait()方法，都可以使当前线程进入 WAITING 等待状态
C. 调用 Object 对象的 notify()方法，可以唤醒 Condition 对象阻塞的线程
D. 调用 Condition 对象的 signalAll()方法，可以唤醒所有正在等待这个条件对象的线程

（3）关于重入锁 ReentrantLock 的描述，正确的是（　　）。
A. ReentrantLock 多次重入调用，容易发生死锁现象
B. ReentrantLock 加锁后，如果忘记了解锁，会发生死锁
C. ReentrantLock 有公平锁和非公平锁模式，默认的模式是公平锁
D. ReentrantLock 是互斥锁，如果忘记了解锁，则其他线程无法再次加锁

（4）关于读锁与写锁，如下描述正确的是（　　）。
A. 读锁是共享锁，写锁是排他锁
B. 读锁与读锁之间是共享的，读锁与写锁之间是互斥的
C. 读锁与读锁之间是共享的，读锁与写锁之间也是共享的
D. 读锁的运算性能比写锁的运算性能高

第 6 章 线程池与阻塞队列

线程池的任务排队需要使用 BlockingQueue 阻塞队列,本章介绍各种常用的阻塞队列。

6.1 Queue 接口

队列是一种特殊的集合,一般队列都具有先进先出(FIFO)的特性(并不绝对要求)。优先级队列(PriorityQueue)按照元素的比较方法排序,其他队列基本采用自然序排队。

```
public interface Queue<E> extends Collection<E> {
    boolean offer(E e);
    E poll();
    E peek();
}
```

队列 Queue 接口实现了 Collection 接口,offer()方法负责把元素插入队列中。peek()方法检索队列头是否存在元素(不移除元素),poll()方法则是检索并移除元素。当队列为空时,peek()和 poll()方法都返回 null。

6.2 BlockingQueue 接口

BlockingQueue 为阻塞队列接口,它继承了 Queue 接口。

```
public interface BlockingQueue<E> extends Queue<E>{
    void put(E e) throws InterruptedException;
    E take() throws InterruptedException;
}
```

BlockingQueue 接口中新增的 put()方法也用于向阻塞队列中插入元素,它与 offer()方法类似,但二者之间也有区别。调用 offer()方法把元素插入队列时,如果队列已满或存在其他限制,则插入操作失败,立即抛出异常信息,方法结束。调用 put()方法把元素插入队列时,如果队列已满,则调用线程会一直阻塞等待,直到队列空间有效,插入数据后这个线程才能返回。

BlockingQueue 接口中新增的 take()方法用于检索并移除队列头元素，它与 poll()方法类似，但二者之间也有区别。调用 poll()方法，如果队列为空，则方法立即返回 null。调用 take()方法，如果队列为空，调用线程会一直阻塞等待，直到队列中的元素有效才会返回。

6.3 BlockingQueue 实现类

BlockingQueue 阻塞队列常用子类具体见表 6-1。

表 6-1　BlockingQueue 实现类

类　名	描　述
ArrayBlockingQueue	基于数组实现的有限阻塞队列，这个队列中的元素先进先出（FIFO）。元素从队列尾部插入，从队列头部移除。队列创建后，"有界缓冲区"的大小不能修改
LinkedBlockingQueue	基于链表实现的阻塞队列，这个队列中的元素先进先出（FIFO）。队列容量可以指定，如果使用默认值，则为 Integer.MAX_VALUE，即"无界缓冲区"
PriorityBlockingQueue	无界阻塞队列，队列中的元素需要实现 Comparable 接口，按照比较顺序排列
SynchronousQueue	同步阻塞队列中不存储任何元素，其容量始终为 0。这是一个数据交换的通道，当插入元素的线程和移除元素的线程单独存在时，需要阻塞等待；成对出现，则完成元素的转移（从插入者移交到获取者）
DelayQueue	一个无界限的阻塞队列，其中元素只能在其延迟到期时才被使用

6.4 LinkedBlockingQueue 与 ArrayBlockingQueue

LinkedBlockingQueue 和 ArrayBlockingQueue 都是先进先出的阻塞队列，它们都具有很好的并发性与线程安全性。

阻塞队列也是集合，它们与传统的 LinkedList 和 ArrayList 集合相似，但是却拥有更好的并发性能。调用 Collections.synchronizedList()方法，可以给 LinkedList 和 ArrayList 带来一定的线程安全性，但这些安全性都是通过简单的给方法增加 synchronized 同步带来的。synchronized 会默认锁定当前对象，同步方法之间互斥，因此 Collections.synchronizedList() 包装后的集合，其并发性能是非常低效的。

ArrayBlockingQueue 队列是一个有界的阻塞队列，底层的数据结构为数组。ArrayBlockingQueue 队列的头部是队列中最先加入的元素。尾部是队列中最后加入的元素。新元素插入队列的尾部，检索操作获取队列头部的元素。有界缓冲区的好处在于可以有效地控制内存的使用，不会出现由于任务过多导致系统崩溃的现象。

```
public class ArrayBlockingQueue<E>
          extends AbstractQueue<E>
          implements BlockingQueue<E>, Serializable{
```

```java
    final Object[] items;
    public ArrayBlockingQueue(int capacity) {
        this(capacity, false);
    }
    public ArrayBlockingQueue(int capacity, boolean fair) {}
    ...
}
```

ArrayBlockingQueue 的缓冲区大小,必须在创建队列时通过构造函数传入,它默认使用非公平锁模式。即排队进入队列的线程,是竞争关系,不遵循公平的排队原则。非公平模式性能优于公平模式,因此除非有特殊要求,尽量使用非公平模式。

LinkedBlockingQueue 底层数据结构为链表。它可以设置为有界,默认空间为 Integer.MAX_VALUE,即"无界缓冲区"。参见下面的源代码,静态内部类 Node 是链表节点。head 是链表头节点,last 是链表的尾节点。

```java
public class LinkedBlockingQueue<E>
        extends AbstractQueue<E>
        implements BlockingQueue<E>, java.io.Serializable {
    transient Node<E> head;
    transient Node<E> last;
    public LinkedBlockingQueue() {
        this(Integer.MAX_VALUE);
    }
    public LinkedBlockingQueue(int capacity) {
    }
    static class Node<E> {
        E item;
        Node<E> next;

        Node(E x) {
            item = x;
        }
    }
    ...  }
```

ArrayBlockingQueue 的任务排队默认为非公平策略,但是它也支持公平策略。而 LinkedBlockingQueue 不支持公平策略,所有任务都是抢占式的。

6.4.1 阻塞队列的单锁与双锁

LinkedBlockingQueue 比 ArrayBlockingQueue 阻塞队列的并发性能更好,吞吐量更大。即当有大量的并发任务同时进出阻塞队列时,应该首选 LinkedBlockingQueue 队列。

阻塞队列的底层线程同步，基本都使用的是重入锁 ReentrantLock。

ArrayBlockingQueue 采用单锁控制，而 LinkedBlockingQueue 采用双锁控制。参见下面的源代码：

```
public class ArrayBlockingQueue<E>
        extends AbstractQueue<E>
        implements BlockingQueue<E>, java.io.Serializable {
    final ReentrantLock lock;
    private final Condition notEmpty;
    private final Condition notFull;
}
public class LinkedBlockingQueue<E>
        extends AbstractQueue<E>
    implements BlockingQueue<E>, java.io.Serializable {
    private final ReentrantLock takeLock = new ReentrantLock();
    private final Condition notEmpty = takeLock.newCondition();
    private final ReentrantLock putLock = new ReentrantLock();
    private final Condition notFull = putLock.newCondition();
}
```

6.4.2 ArrayBlockingQueue 并发分析

ArrayBlockingQueue 的主要源代码如下：

```
public class ArrayBlockingQueue<E>
            extends AbstractQueue<E>
            implements BlockingQueue<E>, java.io.Serializable {
    final ReentrantLock lock;
    private final Condition notEmpty;
    private final Condition notFull;
    public void put(E e) throws InterruptedException {
        lock.lockInterruptibly();
        try {
            while (count == items.length)
                notFull.await();
            enqueue(e);
        } finally {
            lock.unlock();
        }
    }
    public E take() throws InterruptedException {
        lock.lockInterruptibly();
        try {
```

```java
            while (count == 0)
                notEmpty.await();
            return dequeue();
        } finally {
            lock.unlock();
        }
    }
    private void enqueue(E x) {
        items[putIndex] = x;
        if (++putIndex == items.length)
            putIndex = 0;
        count++;
        notEmpty.signal();
    }
    private E dequeue() {
        E x = (E) items[takeIndex];
        items[takeIndex] = null;
        if (++takeIndex == items.length)
            takeIndex = 0;
        count--;
        if (itrs != null)
            itrs.elementDequeued();
        notFull.signal();
        return x;
    }
    ...
}
```

通过源代码分析可以看出，ArrayBlockingQueue 使用了一个重入锁和两个监视器。某个线程想调用 ArrayBlockingQueue 队列写入数据 put() 或读取数据 take() 时，都要先获得同一个 ReentrantLock 锁。因此 put() 方法与 take() 方法具有互斥性。

6.4.3　LinkedBlockingQueue 并发分析

LinkedBlockingQueue 的主要源代码如下：

```java
public class LinkedBlockingQueue<E>
        extends AbstractQueue<E>
        implements BlockingQueue<E>, java.io.Serializable {
    private final ReentrantLock takeLock = new ReentrantLock();
    private final Condition notEmpty = takeLock.newCondition();
    private final ReentrantLock putLock = new ReentrantLock();
    private final Condition notFull = putLock.newCondition();
    public void put(E e) throws InterruptedException {
```

```java
        int c = -1;
        Node<E> node = new Node<E>(e);
        putLock.lockInterruptibly();
        try {
            while (count.get() == capacity) {
                notFull.await();
            }
            enqueue(node);
            c = count.getAndIncrement();
            if (c + 1 < capacity)
                notFull.signal();
        } finally {
            putLock.unlock();
        }
        if (c == 0)
            signalNotEmpty();
    }
    public E take() throws InterruptedException {
        E x;
        int c = -1;
        takeLock.lockInterruptibly();
        try {
            while (count.get() == 0) {
                notEmpty.await();
            }
            x = dequeue();
            c = count.getAndDecrement();
            if (c > 1)
                notEmpty.signal();
        } finally {
            takeLock.unlock();
        }
        if (c == capacity)
            signalNotFull();
        return x;
    }
    private void enqueue(Node<E> node) {
        last = last.next = node;
    }
    private E dequeue() {
        Node<E> h = head;
        Node<E> first = h.next;
        h.next = h;
```

```
            head = first;
            E x = first.item;
            first.item = null;
            return x;
        }
    }
```

通过源代码分析可以看出，LinkedBlockingQueue 使用了两个完全不同的重入锁 (takeLock 和 putLock)，每个重入锁使用一个监视器。当并发读写同一个 LinkedBlockingQueue 队列时，put()方法与 take()方法没有互斥性，互不干扰。

6.4.4 案例：12306 抢票

案例场景：中国铁路 12306 网站，每当春节、国庆等传统节假日来临时，都会有一波抢票高峰。某个车次某天的车票，会提前告知放票时间。大量人员会提前等待，然后在预定时间同时发出购票请求。

在实际的项目开发中，抢票请求会存储在消息队列（MQ）中。这些消息队列都是第三方的中间件，如 ActiveMQ、RabbitMQ、RocketMQ、Kafka 等。这些中间件需要部署在独立的服务器集群中。消息队列的消息入队与出队模式，如图 6-1 所示。

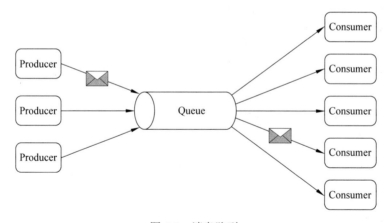

图 6-1　消息队列

使用阻塞队列 BlockingQueue，完全可以模拟消息中间件的抢票和出票场景。
操作步骤如下。

（1）由于抢票人员很多，使用无界队列 LinkedBlockingQueue 比 ArrayBlockingQueue 更加合适。而且 LinkedBlockingQueue 采用双锁模式，抢票请求的入队与出队互不影响，并发性能更高。

```
BlockingQueue<String> mq = new LinkedBlockingQueue<>();
```

（2）使用 newCachedThreadPool 创建线程池，模拟大量的抢票请求。所有抢票信息都存储到消息队列 queue 中（见图 6-1）。由于抢票用户数量不定，所以使用随机数模拟抢票用户数量。

```java
ExecutorService pool = Executors.newCachedThreadPool();
    int requestNum = (int)(Math.random()*5000);
    for (int i = 0; i < requestNum; i++) {
        pool.execute(new Runnable() {
            public void run() {
                try {
                    String msg = Thread.currentThread().getId() + "-购票信息：xxx";
                    mq.put(msg);
                    Thread.sleep(5);
                    System.out.println(msg);
                } catch (Exception e) {
                    e.printStackTrace();
                }
            }
        });
    }
```

（3）使用 newFixedThreadPool 模拟出票处理，处理订票请求的线程数量是固定的（见图 6-1 中的 Consumer）。

```java
ExecutorService pool2 = Executors.newFixedThreadPool(8);
while(true) {
    pool2.execute(new Runnable() {
        public void run() {
            try {
                String msg = mq.take();
                System.out.println("已处理：" + msg);
            } catch (Exception e) {
            }
        }
    });
}
```

（4）程序运行效果如下：

8-购票信息：xxx
10-购票信息：xxx
21-购票信息：xxx
25-购票信息：xxx
17-购票信息：xxx
已处理：13-购票信息：xxx

已处理：14-购票信息：xxx
22-购票信息：xxx
26-购票信息：xxx
24-购票信息：xxx
16-购票信息：xxx
已处理：15-购票信息：xxx
已处理：16-购票信息：xxx
已处理：17-购票信息：xxx
已处理：18-购票信息：xxx
已处理：19-购票信息：xxx
已处理：20-购票信息：xxx
已处理：22-购票信息：xxx
已处理：21-购票信息：xxx
已处理：23-购票信息：xxx
已处理：24-购票信息：xxx
已处理：25-购票信息：xxx
已处理：26-购票信息：xxx
已处理：27-购票信息：xxx
20-购票信息：xxx
已处理：12-购票信息：xxx
12-购票信息：xxx
已处理：11-购票信息：xxx
13-购票信息：xxx
已处理：9-购票信息：xxx
18-购票信息：xxx
9-购票信息：xxx
已处理：10-购票信息：xxx
14-购票信息：xxx
已处理：8-购票信息：xxx
11-购票信息：xxx
15-购票信息：xxx
19-购票信息：xxx
23-购票信息：xxx
27-购票信息：xxx

6.5 生产者与消费者模式

消息队列 MQ（如 ActiveMQ）和阻塞队列（如 LinkedBlockingQueue），元素插入与取出的模式，都是经典的生产者与消费者模式。即生产者写入消息，消费者提取消息。

生产者与消费者是完全解耦的，生产者无须知道消费者是谁；消费者也无须知道生产者是谁。BlockingQueue 对于生产者与消费者的数量没有约束。

虽然生产者-消费者模式可以把入队与出队的代码解耦，但是它们的行为还是间接地通过队列耦合在了一起。它理想地假设消费者将会持续工作，所以你不需要为队列的大小划定边界，但是这将成为日后需要重构系统的隐患。所以在设计初期就使用阻塞队列建立对资源的管理，提早做这件事情会比日后再修复容易得多。在某些情况下，阻塞队列使用更加简单，但是如果阻塞队列并不完全适用于你的设计，你也可以考虑使用 Semaphore 信号灯创建其他的阻塞数据结构。

6.5.1　基于管道发送与接收消息

管道可以作为线程交换数据的通道，PipedWriter 类允许线程向管道中写入数据，PipedReader 类允许线程从管道中读取数据。通过管道读写数据，也是典型的生产者-消费者模式。

管道测试的操作步骤如下。

（1）Sender 为消息的生产者。

```java
class Sender implements Runnable {
    PipedWriter pipedWriter = new PipedWriter();
    public PipedWriter getPipedWriter() {
        return pipedWriter;
    }
    public void run() {
        try {
            while (true) {
                int rand = (int)(Math.random()*100);
                System.out.println(Thread.currentThread().getId()
                        + "向管道写入:" + rand);
                pipedWriter.write(rand);
                Thread.sleep(500);
            }
        } catch (Exception e) { }
    }
}
```

（2）Receiver 是消息的消费者。

```java
class Receiver implements Runnable {
    PipedReader pipedReader = null;

    public Receiver(Sender sender) {
        try {
            pipedReader = new PipedReader(sender.getPipedWriter());
        } catch (Exception e) {
```

```
                e.printStackTrace();
            }
        }
        public void run() {
            try {
                while (true) {
                    int msg = pipedReader.read();
                    System.out.println(Thread.currentThread().getId()
                            + "从管道读取:" + msg);
                }
            } catch (IOException e) {        }
        }
    }
```

（3）在主函数中启动两个线程，分别进行消息的生产与消费。

```
    public static void main(String[] args) {
        try {
            Sender sender = new Sender();
            Receiver reciever = new Receiver(sender);
            new Thread(sender).start();
            new Thread(reciever).start();
        } catch (Exception e) {
            e.printStackTrace();
        }
    }
```

程序运行结果如下：

8 向管道写入:61
8 向管道写入:20
9 从管道读取:61
9 从管道读取:20
8 向管道写入:20
8 向管道写入:60
9 从管道读取:20
9 从管道读取:60
8 向管道写入:98
8 向管道写入:65
9 从管道读取:98
9 从管道读取:65

6.5.2 基于阻塞队列发送与接收消息

4.4.3 节的日志读写案例，5.2.2 节的缓冲区队列案例和 6.4.4 节的 12306 抢票案例，都是典型的基于阻塞队列的生产者与消费者模式。

6.5.3 案例：医院挂号

案例场景描述：病人挂号后，医生才会叫号就诊，否则医生并不知道接下来该给谁看病。这是一个典型的生产者和消费者场景，其中病人充当了生产者，只有当病人产生了挂号单，消费者（也就是医生）才会去处理这个挂号单。

操作步骤如下：

(1) 新建挂号单类 Order。

```java
class Order {
    private int id;           //挂号单序号
    public int getId() {
        return id;
    }
    public Order(int id) {
        this.id = id;
    }
}
```

(2) 新建阻塞队列，用于存挂号单。

```java
final BlockingQueue<Order> queue = new LinkedBlockingQueue<Order>();
```

(3) 使用独立线程，模拟窗口挂号过程。

```java
new Thread(new Runnable() {
    public void run() {
        for(int i=0;i<50;i++) {
            try {
                //模拟挂号过程
                Thread.sleep(300);
                Order order = new Order(i);
                queue.put(order);
                System.out.println(i + "号已挂...");
            } catch (Exception e) {
                e.printStackTrace();
            }
        }
         System.out.println("今天的号挂完了，请明天再来...");
```

```
        }
}).start();
```

(4)使用线程池,模拟医生看病过程。

```
ExecutorService pool = Executors.newFixedThreadPool(3);
    for(int i=0;i<3;i++) {
     pool.execute(new Runnable() {
             public void run() {
                 while(true) {
                    try {
                        Order order = (Order)queue.take();
                        int rand = (int)(Math.random()*3000);
                        Thread.sleep(rand);
                        System.out.println("医生: " +
                           Thread.currentThread().getId()
                               + ",给" + order.getId() + "号病人看完了...");
                    } catch (Exception e) {
                        e.printStackTrace();
                    }
                }
            }
        });
    }
pool.shutdown();
```

程序运行结果如下:

0 号已挂...
1 号已挂...
2 号已挂...
医生:9,给 2 号病人看完了...
3 号已挂...
4 号已挂...
5 号已挂...
医生:9,给 3 号病人看完了...
6 号已挂...
7 号已挂...
医生:11,给 1 号病人看完了...
8 号已挂...
医生:10,给 0 号病人看完了...
9 号已挂...
10 号已挂...
11 号已挂...

医生：9,给 4 号病人看完了...
12 号已挂...
13 号已挂...
医生：11,给 5 号病人看完了...
14 号已挂...
15 号已挂...
16 号已挂...
17 号已挂...
医生：10,给 6 号病人看完了...
18 号已挂...
19 号已挂...
医生：9,给 7 号病人看完了...
20 号已挂...
医生：11,给 8 号病人看完了...
21 号已挂...
医生：11,给 11 号病人看完了...
22 号已挂...
23 号已挂...
24 号已挂...
医生：9,给 10 号病人看完了...
25 号已挂...
医生：10,给 9 号病人看完了...
26 号已挂...
27 号已挂...
28 号已挂...
医生：11,给 12 号病人看完了...
29 号已挂...
30 号已挂...
医生：10,给 14 号病人看完了...
31 号已挂...
32 号已挂...
33 号已挂...
医生：9,给 13 号病人看完了...
34 号已挂...
35 号已挂...
医生：11,给 15 号病人看完了...
36 号已挂...
37 号已挂...
医生：9,给 17 号病人看完了...
医生：10,给 16 号病人看完了...
38 号已挂...
39 号已挂...
40 号已挂...

医生：9,给19号病人看完了...
41号已挂...
医生：10,给20号病人看完了...
42号已挂...
43号已挂...
医生：11,给18号病人看完了...
44号已挂...
医生：9,给21号病人看完了...
45号已挂...
46号已挂...
47号已挂...
医生：10,给22号病人看完了...
48号已挂...
49号已挂...
今天的号挂完了，请明天再来...
医生：11,给23号病人看完了...
医生：10,给25号病人看完了...
医生：9,给24号病人看完了...
医生：11,给26号病人看完了...
医生：10,给27号病人看完了...
医生：11,给29号病人看完了...
医生：10,给30号病人看完了...
医生：11,给31号病人看完了...
医生：9,给28号病人看完了...
医生：9,给34号病人看完了...
医生：11,给33号病人看完了...
医生：10,给32号病人看完了...
医生：11,给36号病人看完了...
医生：10,给37号病人看完了...
医生：9,给35号病人看完了...
医生：11,给38号病人看完了...
医生：10,给39号病人看完了...
医生：9,给40号病人看完了...
医生：10,给42号病人看完了...
医生：9,给43号病人看完了...
医生：11,给41号病人看完了...
医生：9,给45号病人看完了...
医生：11,给46号病人看完了...
医生：9,给47号病人看完了...
医生：10,给44号病人看完了...
医生：11,给48号病人看完了...
医生：9,给49号病人看完了...

6.6 SynchronousQueue

SynchronousQueue 同步队列实现了 BlockingQueue 接口，它是一种非常特殊的队列，因为 SynchronousQueue 队列中没有内部容量，也就是说该队列中不能存储元素，它的容量 size 值始终为 0。

```
public class SynchronousQueue<E>
        extends AbstractQueue<E>
        implements BlockingQueue<E>, java.io.Serializable {}
```

SynchronousQueue 没有内部容量，其作用显然不是用于元素存储，它的主要用途类似 CSP 和 Ada 中使用的会合通道，即这是一个元素数据交换的通道。只有等待插入元素的线程与等待提取元素的线程同时存在时，才会利用 SynchronousQueue 通道完成数据的交换。如果插入线程与提取线程只有一方存在，SynchronousQueue 会起到线程阻塞等待的作用。

你不能在同步队列中使用 peek()方法，因为一个元素，当你尝试删除它时才存在。 你无法插入元素（使用任何方法），除非另有线程正在尝试删除它。你不能迭代，因为队列中没有什么元素可以迭代。队列的头部是第一个排队等待插入的线程，它就是准备添加到队列中的元素。如果没有排队线程，那么没有元素可用于删除，并且 poll()方法将返回 null。为了使父接口 Collection 的方法可以使用，SynchronousQueue 充当空集合，但是同步队列中不允许使用 null 元素。

相比于其他阻塞队列，如 ArrayBlockingQueue 和 LinkedBlockingQueue 等阻塞队列，SynchronousQueue 是一个轻量级队列，它非常直接地移交元素，减少了生产者和消费者之间移动数据的延迟时间。SynchronousQueue 没有存储能力，因此除非另一个线程已经准备好参与移交工作，否则 put()方法和 take()方法将被一直阻塞。

6.6.1 同步队列应用场景

SynchronousQueue 同步阻塞队列只有在消费者数量充足的时候才适合使用，因为队列没有存储空间，这样可以为下一个任务做好充分准备。

Executors 工具类中的 newCachedThreadPool 底层就是使用的 SynchronousQueue 创建的同步队列。利用 SynchronousQueue 的任务交换通道作用，所有向这个线程池请求执行的任务，都可以快速找到执行线程。

```
public static ExecutorService newCachedThreadPool() {
    return new ThreadPoolExecutor(0, Integer.MAX_VALUE,
                                  60L, TimeUnit.SECONDS,
                                  new SynchronousQueue<Runnable>());
}
```

此类支持排队线程的公平策略，对于生产者和消费者线程，均可使用公平策略进行 FIFO 排队访问，但是默认的非公平策略性能更优。

6.6.2 案例：Web 服务器处理并发请求

案例场景描述：客户端浏览器发送 HTTP 请求给 Web 服务器。Web 服务器需要马上对客户端请求进行回应。当有大量的 HTTP 并发请求同时发出时，Web 服务器也不应有任何延缓，而是尽可能快地全部满足客户端请求。

操作步骤如下：

（1）创建 HTTP 请求类，模拟客户端请求。

```java
class HttpRequest{
    private String target;
    public String getTarget() {
        return target;
    }
    public HttpRequest(String target) {
        this.target = target;
    }
}
```

（2）使用 SynchronousQueue，创建 HTTP 请求的任务队列。

```java
BlockingQueue<HttpRequest> taskQueue = new SynchronousQueue<>();
```

（3）模拟客户端的大量并发请求，所有请求信息写入 SynchronousQueue 队列中。

```java
int clientNum = (int)(Math.random()*100);
ExecutorService clientPool = Executors.newFixedThreadPool(clientNum);
for(int i=0;i<clientNum;i++) {
    clientPool.execute(new Runnable() {
        public void run() {
            int target = (int)(Math.random()*10000);
            HttpRequest request = new HttpRequest("http://xxx/" + target);
            try {
                taskQueue.put(request);
                System.out.println("客户-"+ Thread.currentThread().getId()
                        + "请求：" + request.getTarget());
            } catch (InterruptedException e) {
                e.printStackTrace();
            }
        }
    });
}
```

(4) 模拟 Web 服务器，从 SynchronousQueue 中提取请求，然后进行并发处理。

```java
final int coreSize = 5;
ExecutorService serverPool = Executors.newFixedThreadPool(coreSize);
for(int i=0;i<coreSize;i++) {
    serverPool.execute(new Runnable() {
        public void run() {
            while(true) {
                try {
                    HttpRequest request = (HttpRequest)taskQueue.take();
                    System.out.println("服务器线程-"
                                + Thread.currentThread().getId()
                                + "处理: " + request.getTarget() );
                    Thread.sleep(100);
                } catch (Exception e) {
                    e.printStackTrace();
                }
            }
        }
    });
}
```

程序运行结果如下，分析运行结果可以明显地看出，所有客户端请求与服务器处理都是成对出现的（由于不同线程输出速度不一致，部分显示会有先后顺序的出入）：

```
客户-15 请求: http://xxx/2578
服务器线程-24 处理: http://xxx/2578
客户-8 请求: http://xxx/1474
服务器线程-25 处理: http://xxx/1474
客户-20 请求: http://xxx/4492
客户-11 请求: http://xxx/4514
服务器线程-28 处理: http://xxx/4492
服务器线程-26 处理: http://xxx/4514
服务器线程-27 处理: http://xxx/3183
客户-19 请求: http://xxx/3183
服务器线程-24 处理: http://xxx/8217
客户-17 请求: http://xxx/8452
服务器线程-25 处理: http://xxx/8452
客户-13 请求: http://xxx/3324
服务器线程-26 处理: http://xxx/3324
客户-18 请求: http://xxx/1286
服务器线程-28 处理: http://xxx/1286
客户-23 请求: http://xxx/8217
```

客户-9 请求：`http://xxx/4641`
服务器线程-27 处理：`http://xxx/4641`
客户-16 请求：`http://xxx/4124`
服务器线程-28 处理：`http://xxx/4124`
客户-14 请求：`http://xxx/3883`
服务器线程-26 处理：`http://xxx/3883`
客户-12 请求：`http://xxx/532`
服务器线程-25 处理：`http://xxx/532`
客户-22 请求：`http://xxx/6732`
服务器线程-24 处理：`http://xxx/6732`
服务器线程-27 处理：`http://xxx/458`
客户-10 请求：`http://xxx/458`
客户-21 请求：`http://xxx/6452`
服务器线程-28 处理：`http://xxx/6452`

6.7 延迟阻塞队列

DelayQueue 是一个无界的阻塞队列，队列中的所有元素必须要实现 Delayed 接口，而且元素只能在其延迟到期时才能使用。

```
public class DelayQueue<E extends Delayed>
                extends AbstractQueue<E>
                implements BlockingQueue<E>{
}
public interface Delayed
          extends Comparable<Delayed>{
    long getDelay(TimeUnit unit);
}
```

DelayQueue 队列的头部是延迟期满后保存时间最长的 Delayed 元素。如果没有任何延迟到期，那么就不会有任何头元素，并且 poll() 方法将返回 null 元素（队列中不允许添加 null 对象）。

DelayQueue 队列在我们系统开发中也会经常使用到。例如，设计一个缓存系统，缓存中的对象超过了空闲时间，需要从缓存中移出。任务调度系统，能够准确地把握任务的执行时间等。

为了具有调用行为，存放到 DelayQueue 的元素必须实现 Delayed 接口。Delayed 接口使对象成为延迟对象，它使存放在 DelayQueue 队列中的对象具有了激活日期。Delayed 接口有一个名为 getDelay() 的方法，它可以用来告知距离延迟到期有多长时间，或者延迟在多长时间之前已经到期。这个方法将强制使用 TimeUnit 类，因为这就是参数类型。这会产生一个非常方便的类，使用该类可以很容易地转换单位而无须做任何声明。为了排序，Delayed 接口还继承了 Comparable 接口，因此必须实现 compareTo()方法，使其可以产生合理的比较。

6.7.1 案例：元素延迟出队

案例场景描述：提取元素的线程阻塞等待，当某个元素延迟期满后将被移出队列。操作步骤如下。

（1）新建元素类 Bean，实现 Delayed 接口。

```java
class Bean implements Delayed {
    private int id;
    private long beginTime;
    private long delayTime;

    public Bean(long delayTime, int id) {
        this.beginTime = System.currentTimeMillis();
        this.delayTime = delayTime;
        this.id = id;
    }
    public int compareTo(Delayed o) {
        long result = this.getDelay(TimeUnit.MILLISECONDS)
                -o.getDelay(TimeUnit.MILLISECONDS);
        return (int)result;
    }
    public long getDelay(TimeUnit unit) {
        long last = beginTime + this.delayTime - System.currentTimeMillis();
        return last;
    }
    public String toString() {
        return "id-" + id + ",延迟" + delayTime + "毫秒...";
    }
}
```

（2）在主函数中创建延时阻塞队列。

```java
final DelayQueue<Bean> delayQueue = new DelayQueue<Bean>();
```

（3）元素入队，每个元素都设置了延时时间。

```java
new Thread(new Runnable() {
    public void run() {
        for(int i=0;i<20;i++) {
            int delay = (int)(Math.random()*5000);
            delayQueue.put(new Bean(delay, i));
        }
```

        }
    }).start();
```

(4)创建独立线程,从延时阻塞队列中提取数据。

```java
new Thread(new Runnable() {
    public void run() {
        while (true) {
            try {
                Bean bean = (Bean) delayQueue.take();
                System.out.println("出队:" + bean.toString());
            } catch (Exception e) {
                e.printStackTrace();
            }
        }
    }
}).start();
```

程序运行结果如下,延时短的元素,会优先移出队列:

```
出队:id-4,延迟 121 毫秒...
出队:id-19,延迟 131 毫秒...
出队:id-3,延迟 571 毫秒...
出队:id-8,延迟 615 毫秒...
出队:id-14,延迟 727 毫秒...
出队:id-1,延迟 1068 毫秒...
出队:id-9,延迟 1220 毫秒...
出队:id-18,延迟 1298 毫秒...
出队:id-2,延迟 1695 毫秒...
出队:id-6,延迟 1996 毫秒...
出队:id-17,延迟 2096 毫秒...
出队:id-0,延迟 2848 毫秒...
出队:id-5,延迟 3399 毫秒...
出队:id-10,延迟 3587 毫秒...
出队:id-13,延迟 3904 毫秒...
出队:id-15,延迟 4066 毫秒...
出队:id-12,延迟 4267 毫秒...
出队:id-16,延迟 4486 毫秒...
出队:id-7,延迟 4708 毫秒...
出队:id-11,延迟 4779 毫秒...
```

6.7.2 项目案例:Web 服务器会话管理

案例场景描述:在 Web 服务器中,每个客户端的访问都采用会话的方式进行管理。例
```

如，用户进行了登录，就可以在会话对象中存储登录成功后的信息，后面的 HTTP 请求就无须再次进行登录操作了。每个会话 HTTPSession 对象，都有一个 lastAccessedTime 属性。客户在会话期内的每次请求，都会更新 lastAccessedTime 值，然后开启倒计时，在指定时间内没有再次更新 lastAccessedTime，则会话结束。

下面，使用 DelayQueue 模拟 Web 服务器的会话倒计时管理机制，操作步骤如下：

（1）创建会话类 HTTPSession，每个会话表示一个客户端与服务器之间的一次通话，它允许会话有效期内的多次 HTTP 请求。会话期内的每次新的 HTTP 请求，都需要更新 lastAccessedTime 属性值。maxInactiveInterval 属性为两次 HTTP 请求的最大间隔时间，倒计时超过 maxInactiveInterval 设置，没有新的 HTTP 请求，则删除该会话对象。在真实项目中，maxInactiveInterval 默认值为 15 分钟，以下程序中修改为 15 秒。

```java
class HTTPSession {
    private long lastAccessedTime;
    private int maxInactiveInterval;
    private String sessionid;
    public String getSessionid() {
        return sessionid;
    }
    public HTTPSession(String sessionid) {
        this.sessionid = sessionid;
        this.lastAccessedTime = System.currentTimeMillis();
        this.maxInactiveInterval = 15*1000;   //设置超时时间为 15 秒
    }
    public long getLastAccessedTime() {
        return lastAccessedTime;
    }
    public void setLastAccessedTime(long lastAccessedTime) {
        this.lastAccessedTime = lastAccessedTime;
    }
    public int getMaxInactiveInterval() {
        return maxInactiveInterval;
    }
    public void setMaxInactiveInterval(int interval) {
        this.maxInactiveInterval = interval;
    }
    public void invalid() {
        System.out.println(this.sessionid + "会话被销毁...");
    }
}
```

（2）创建类 SessionTask，实现 Delayed 接口，这个类的实例将被用于添加到监控队列中。

```java
class SessionTask implements Delayed {
    private HTTPSession session;
    public HTTPSession getSession() {
        return session;
    }
    public SessionTask(HTTPSession session) {
        this.session = session;
    }
    public int compareTo(Delayed o) {
        long result = this.getDelay(TimeUnit.MILLISECONDS)
                    -o.getDelay(TimeUnit.MILLISECONDS);
        return (int)result;
    }
    public long getDelay(TimeUnit unit) {
        long last = this.session.getLastAccessedTime()
                + this.session.getMaxInactiveInterval()
                - System.currentTimeMillis();
        return last;
    }
}
```

（3）在主函数中创建会话监控队列。

```java
final DelayQueue<SessionTask> delayQueue = new DelayQueue<SessionTask>();
```

（4）在主函数中创建会话集合，用于存储所有会话对象。注意此处不能使用 HashMap，而应该使用有并发控制的 ConcurrentHashMap，HashMap 在并发操作中会抛出异常。

```java
Map<String,HTTPSession> allClients = new ConcurrentHashMap<>();
```

（5）模拟多用户并发访问。第一次 HTTP 请求就需要把 SessionTask 加入延时队列中。如果会话期内有两次、三次 HTTP 请求，需要修改 lastAccessedTime 属性。这个属性值修改后，延时队列中的元素会重新计算延迟时间。

```java
ExecutorService pool = Executors.newCachedThreadPool();
for(int i=0;i<10;i++) {
    pool.execute(new Runnable() {
        public void run() {
            String sessionid = Thread.currentThread().getId()
                            + "-" + System.currentTimeMillis();
            HTTPSession session = new HTTPSession(sessionid);
            allClients.put(sessionid, session);
```

```java
            SessionTask st = new SessionTask(session);
            delayQueue.put(st);   //加入延时移除队列中
            // 模拟二次 HTTP 请求,更新 HTTPSession 的最后访问时间
            int rand = (int)(Math.random()*10);
            if(rand>0) {
                int delay = rand * 1000;
                System.out.println(session.getSessionid()
                            + "延时" + delay + "毫秒后,二次请求...");
                try {
                    Thread.sleep(delay);
                } catch (Exception e) {
                }
                session.setLastAccessedTime(System.currentTimeMillis());
            }
        }
    });
}
```

(6) 使用独立的线程监控会话对象是否超时,已经超时的会话,设置为失效。

```java
ExecutorService monitor = Executors.newSingleThreadExecutor();
monitor.execute(new Runnable() {
    public void run() {
        while(true) {
            try {
                SessionTask st = delayQueue.take();
                String sessionid = st.getSession().getSessionid();
                System.out.println(sessionid + "超时");
                HTTPSession session = allClients.get(sessionid);
                session.invalid();
            } catch (InterruptedException e) {
                e.printStackTrace();
            }
        }
    }
});
```

程序运行结果如下,按照最大间隔时间和 HTTP 的请求时间,真实模拟了 Web 服务器如何管理会话对象,程序运行结果与 Tomcat 等真实 Web 服务器的管理效果基本一致。

```
8-1598167719748 延时 4000 毫秒后,二次请求...
16-1598167719753 延时 3000 毫秒后,二次请求...
9-1598167719748 延时 2000 毫秒后,二次请求...
12-1598167719753 延时 5000 毫秒后,二次请求...
```

```
14-1598167719750 延时 4000 毫秒后,二次请求...
15-1598167719751 延时 6000 毫秒后,二次请求...
13-1598167719751 超时
13-1598167719751 会话被销毁...
17-1598167719751 超时
17-1598167719751 会话被销毁...
10-1598167719748 超时
10-1598167719748 会话被销毁...
11-1598167719750 超时
11-1598167719750 会话被销毁...
9-1598167719748 超时
9-1598167719748 会话被销毁...
16-1598167719753 超时
16-1598167719753 会话被销毁...
8-1598167719748 超时
8-1598167719748 会话被销毁...
14-1598167719750 超时
14-1598167719750 会话被销毁...
12-1598167719753 超时
12-1598167719753 会话被销毁...
15-1598167719751 超时
15-1598167719751 会话被销毁...
```

6.8 PriorityBlockingQueue

PriorityBlockingQueue 是一个无界的优先级队列,队列中的元素需要实现 Comparable 接口。从队列中提取元素时,会按照比较排序的结果顺序提取。

虽然这个队列在逻辑上是无界的,但由于资源耗尽,尝试的添加可能会失败(导致 OutOfMemoryError)。该队列中不允许使用 null 元素,也不允许插入无法比较排序的对象。PriorityBlockingQueue 的操作不会保证相同优先级的元素排序,如果需要强制执行排序,你可以自定义类或比较器,使用辅助键来破坏主优先级值的关系。

案例:按优先级执行任务

案例场景描述:按照任务的优先级排队执行任务,Task 对象中的 priority 越小,表示其优先级越高。

操作步骤如下:

(1)新建任务 Task,实现 Comparable 接口。按照优先级进行任务排序。

```
class Task implements Comparable<Task> {
    private int priority;
```

```java
    private String tno;
    public int getPriority() {
        return priority;
    }
    public String getTno() {
        return tno;
    }
    public Task(int priority, String tno) {
        this.priority = priority;
        this.tno = tno;
    }
    public int compareTo(Task o) {
        return this.priority - o.priority;
    }
}
```

（2）在主函数中创建优先级队列。

```java
final BlockingQueue<Task> queue = new PriorityBlockingQueue<Task>();
```

（3）模拟多个任务，并发写入优先级队列中。

```java
ExecutorService pool = Executors.newFixedThreadPool(10);
for (int i = 0; i < 10; i++) {
    pool.execute(new Runnable() {
        public void run() {
            int priority = (int) (Math.random() * 100);
            String tno = "t" + Thread.currentThread().getId();
            Task task = new Task(priority,tno);
            try {
                queue.put(task);
            } catch (Exception e) {
                e.printStackTrace();
            }
        }
    });
}
```

（4）使用独立线程，提取优先级队列中的任务。

```java
new Thread() {
    public void run() {
        while (true) {
            try {
                Task task = (Task) queue.take();
```

```
                        System.out.println("执行优先级为" + task.getPriority()
                                    + "的任务:" + task.getTno());
                } catch (Exception e) {
                    e.printStackTrace();
                }
            }
        }
    }
}.start();
```

程序运行结果如下,优先级别高的任务会被优先执行:

执行优先级为 7 的任务:t12
执行优先级为 30 的任务:t15
执行优先级为 35 的任务:t11
执行优先级为 61 的任务:t10
执行优先级为 62 的任务:t9
执行优先级为 67 的任务:t17
执行优先级为 76 的任务:t8
执行优先级为 80 的任务:t13
执行优先级为 90 的任务:t16
执行优先级为 92 的任务:t14

6.9 LinkedTransferQueue

LinkedTransferQueue 是一个基于链接节点的无界阻塞队列,所有元素满足 FIFO(先进先出),队列的头部是那些最早排队的生产者, 队列的尾部则是那些最后入队的生产者。相对于其他阻塞队列,LinkedTransferQueue 多了 transfer()方法和 tryTransfer()方法。

```
public class LinkedTransferQueue<E>
            extends AbstractQueue<E>
            implements TransferQueue<E>, Serializable{
    public int remainingCapacity() {}
    public boolean tryTransfer(E e) {}
    public void transfer(E e) throws InterruptedException {}
    public boolean tryTransfer(E e, long timeout, TimeUnit unit)
                            throws InterruptedException {}
    public boolean hasWaitingConsumer() {}
    public int getWaitingConsumerCount() {}
}
```

Transfer()方法:如果当前有消费者正在等待接收元素,transfer()方法可以把生产者传入的元素立刻传输给消费者。如果没有消费者在等待接收元素,transfer()方法会将元素插入该队列的尾部，并等待消费者接收该元素。

tryRransfer()方法：如果存在已经等待接收元素的消费者，会立即转移指定的元素，否则将指定的元素插入该队列尾部，并等待消费者接收该元素。如果可以在指定时间之前将元素传送给消费者，则返回 true，否则返回 false。

相对于 SynchronousQueue 这种直来直去传输的阻塞队列，LinkedTransferQueue 多了一个存储空间。相对于 LinkedBlockingQueue 阻塞队列，LinkedTransferQueue 多了一个直接将元素传输给消费者的功能。在某些场合下，LinkedTransferQueue 队列的性能更高，用途也更灵活多变。

6.10　LinkedBlockingDeque

LinkedBlockingDeque 是一个由链接节点组成的双向阻塞队列。

```
public class LinkedBlockingDeque<E>
        extends AbstractQueue<E>
        implements BlockingDeque<E>, Serializable {
    public void put(E e) {}
    public void putFirst(E e) {}
    public void putLast(E e){}
    public E take() {}
    public E takeFirst(){}
    public E takeLast(){}
    ...
}
```

所谓双向队列指的是从队列的两端可以同时插入和移出元素，因此在高并发的入队与出队操作时，其性能更佳。

与 LinkedBlockingQueue 单向队列相比，LinkedBlockingDeque 增加了 addFirst()、addLast()、offerFirst()、offerLast()、peekFirst()、peekLast()、putFirst()、putLast()、takeFirst()、takeLast()等方法。

LinkedBlockingDeque 默认为无界阻塞队列，但有时为了防止其容量过度扩展，可以在初始化队列时设置它的最大容量。

6.11　本章习题

（1）如下关于阻塞队列的描述，正确的是（　　）。
A．调用 put()方法把元素插入阻塞队列时，如果队列已满，线程会一直阻塞等待
B．调用 offer()方法把元素插入阻塞队列时，如果队列已满，线程会一直阻塞等待
C．调用 take()方法移除元素时，如果阻塞队列为空，调用线程会一直阻塞等待

D. 调用 poll() 方法移除元素时，如果阻塞队列为空，调用线程会一直阻塞等待

（2）如下关于阻塞队列的描述，正确的是（　　）。

A. ArrayBlockingQueue 和 LinkedBlockingQueue 都是有界队列

B. SynchronousQueue 队列中不能存储任何元素，它的 size 始终为 0

C. ArrayBlockingQueue 是有界队列，LinkedBlockingQueue 是无界队列

D. PriorityBlockingQueue 中的元素，按照元素插入的自然序排列

（3）如下关于 LinkedBlockingQueue 和 ArrayBlockingQueue 的描述，不正确的是（　　）。

A. ArrayBlockingQueue 队列是一个有界的阻塞队列，底层的数据结构为数组；LinkedBlockingQueue 底层数据结构为链表，它是无界阻塞队列

B. LinkedBlockingQueue 和 ArrayBlockingQueue 都是先进先出的阻塞队列，它们都具有很好的并发性与线程安全性

C. LinkedBlockingQueue 与 ArrayBlockingQueue 阻塞等待读写数据时，被阻塞线程的状态都为 BLOCKED

D. ArrayBlockingQueue 底层同步采用一个重入锁，LinkedBlockingQueue 底层使用两个重入锁，LinkedBlockingQueue 的并发读写性能更好

（4）如下关于生产者与消费者模式的描述，不正确的是（　　）。

A. 阻塞队列的元素插入与读取，是经典的生产者与消费者模式

B. 生产者与消费者是完全解耦的，生产者无须知道消费者是谁，消费者也无须知道生产者是谁

C. 使用 LinkedBlockingQueue 作为阻塞队列，生产者与消费者是互不干扰的

D. 使用 ArrayBlockingQueue 作为阻塞队列，生产者与消费者也可以互不干扰

第 7 章 线程池与 AQS

java.util.concurrent 包中的绝大多数同步工具，如锁（locks）和屏障（barriers）等，都基于 AbstractQueuedSynchronizer（简称 AQS）构建而成。这个框架提供了一套同步管理的通用机制，如同步状态的原子性管理、线程阻塞与解除阻塞，还有线程排队等。

在 JDK1.5 引入了 java.util.concurrent 包，其中包含多个支持中等级别线程并发的类，如可重入锁（ReentrantLock）、读锁（ReentrantReadWriteLock.ReadLock）、写锁（ReentrantReadWriteLock.WriteLock）、信号量（Semaphore）、屏障（CyclicBarrier）、Future 对象、事件指示器以及传送队列等。这些同步类主要有如下功能：

（1）对象内部同步状态的维护（如表示锁的状态是已获取还是已释放）。

（2）更新和检查状态的操作。而且至少有一个方法会导致调用线程在同步状态被获取时被阻塞，以及在其他线程改变这个同步状态时解除线程的阻塞。

几乎任何一个知名的同步器都可以用来实现其他形式的同步器。例如，可以用可重入锁（ReentrantLock）来实现信号量（Semaphore）；反之，用信号量也可以实现可重入锁。但是，这样做会带来复杂性高、开销过大、不灵活等问题，使其最终只能成为一个二流项目。而使用 AQS 用户可以用简洁的方式定义自己的线程同步器。

7.1 acquire 与 release

所有的线程同步器至少应该包含两个方法：一个是 acquire，另一个是 release。acquire 方法阻塞调用的线程，直到同步状态允许其继续执行。而 release 操作则是通过某种方式改变同步状态，使得一个或多个被阻塞的线程解锁。

```
public abstract class AbstractQueuedSynchronizer
                extends AbstractOwnableSynchronizer
                implements java.io.Serializable {
    public final void acquire(int arg) {}
    public final boolean release(int arg) {}
    protected boolean tryAcquire(int arg) {}
    protected boolean tryRelease(int arg) {}
```

}

在 JUC 包中并没有对同步器定义统一的 API。因此，有些类通过 Lock 接口来定义（如 ReentrantLock、ReentrantReadWriteLock.ReadLock、ReentrantReadWriteLock.WriteLock），而另外一些则定义了其专有的版本（如 Semaphore、CountDownLatch）。因此 acquire 和 release 方法的操作，会有各种不同的形式、不同的类。例如，Lock.lock、Semaphore.acquire、CountDownLatch.await 和 FutureTask.get 等，都会映射到 acquire 操作。

JUC 为了支持同步器的常用功能，对所有同步类做了一致性约定，即每个同步器都应支持下面的操作：

- 阻塞和非阻塞同步尝试（如 tryLock）。
- 可选的超时设置，让调用者可以放弃等待。
- 通过中断取消任务，通常会分为两个版本：一个可以取消，而另一个不可以。

同步器的实现根据其状态是否独占而有所不同。独占状态的同步器，在同一时间只有一个线程可以通过阻塞点，而共享状态的同步器可以同时有多个线程进入阻塞点。通过 AQS 实现的同步器，必须同时支持独占与共享两种模式。

在 JUC 包里还定义了 Condition 接口，支持监视器风格的 await/signal 操作。Condition 的操作必须与独占模式的 Lock 类相关。

7.2　性能目标

Java 内置锁（使用 synchronized 的方法或代码块）的性能问题一直以来都被人们关注，并且已经有一系列的文章描述了其构造。然而，大部分研究主要关注的是在单核处理器上用于单线程上下文环境中时，如何尽量降低其空间（因为任何 Java 对象都可以服务于锁）和时间的开销。

对于同步器来说这些都不是特别重要：程序员仅在需要的时候才会使用同步器，因此并不需要压缩空间来避免浪费；而且同步器都是专门用于多线程设计中（主要是多核处理器），在这种环境下，偶尔的 CPU 竞争是完全正常可预期的，无须过度担心。

synchronized 的监视器锁依赖于 JVM，常规的 JVM 锁优化策略用于避免 CPU 竞争。这种优化策略对于依赖多核服务器的 JUC 同步器来说，并不适用。

本节讨论的主要性能目标是可伸缩性，即在大部分情况下，特别是在同步器有竞争的情况下，稳定地保证其效率。在理想的情况下，不管有多少线程正试图通过阻塞点，通过阻塞点的开销都应该是个常量。在某一线程被允许通过阻塞点但还没有通过的情况下，使其耗费的总时间最少，这是主要目标之一。当然，必须要考虑平衡各种资源，如 CPU 时间的总消耗、内存竞争以及线程调度的开销。例如，自旋锁通常比阻塞锁 acquire 所需的时间更短，但是通常也会浪费 CPU 时钟周期，并且造成内存竞争，所以使用得并不多。

对于那些需要控制资源分配的应用，更重要的是去维持多线程读取的公平性，可以容忍较差的吞吐量。因此，应该提供不同的公平策略。

无论同步器的内部实现多么精致，它还是会在某些应用中产生性能瓶颈。因此，AQS
必须提供监视工具让用户发现这些瓶颈，至少需要提供一种方式来确定有多少线程被阻塞了。

7.3 设计与实现

实现 acquire 方法的思路如下：

```
while (同步状态不允许 acquire) {
    入队当前线程
    阻塞当前线程
}
出队当前线程
```

实现 release 方法的思路如下：

```
更新同步状态；
if (状态允许某个阻塞线程 acquire){
    解锁一个或多个已入队的线程
}
```

为了实现上述操作，需要三个基本组件的相互协作：同步状态的原子性管理、阻塞与解阻塞线程、线程排队管理。虽然上述三个组件的实现可以是独立的，但这样会导致难用和没有效率，因此需要三个组件协同一起工作。例如，存储队列节点的信息必须与解除阻塞所需要的信息一致，而暴露出的方法签名必须依赖于同步状态的特性。

AQS 有意限制了其适用范围，同时提供了足够高的效率。因此，除非有特殊需求，使用 AQS 作为同步器比你重新构建一个同步器要好得多。

7.3.1 同步状态

AQS 类使用一个 int（32 位）值来保存同步状态，并暴露出 getState、setState 以及 compareAndSetState 操作来读取和更新这个状态。这些方法都依赖于 java.util.concurrent.atomic 包支持，这个包提供了对 volatile 关键字在读和写语义上的支持，并且通过使用本地的 compare-and-swap 或 load-linked/store-conditional 指令来实现原子性的数据比较赋值，只有同步状态拥有一个期望值的时候，才会被原子地设置成新值。

```
public abstract class AbstractQueuedSynchronizer
            extends AbstractOwnableSynchronizer
            implements java.io.Serializable {
    private volatile int state;
    protected final int getState() {
        return state;
    }
```

```java
    protected final void setState(int newState) {
        state = newState;
    }
    protected final boolean compareAndSetState(int expect, int update) {
        return unsafe.compareAndSwapInt(this, stateOffset, expect, update);
    }
}
```

将同步状态限制为一个 32 位的整形是出于实践上的考量。虽然 JSR166 也提供了 64 位 long 字段的原子性操作，但这些操作在很多平台上还是使用内部锁的方式来模拟实现的，这可能会使同步器的性能不理想。目前来说，32 位的状态对大多数应用程序都是足够的。在 JUC 包中，只有一个同步器类可能需要多于 32 位来维持状态，那就是 CyclicBarrier 类，因此它必须调用 tryAcquire 和 tryRelease 方法来设置同步状态。

1. volatile 关键字

volatile 是一个特征修饰符，它的作用是作为指令关键字，确保本条指令不会因编译器的优化而省略，而且要求直接读取这个变量值，不能从缓存或寄存器等位置读取。

如下所示的 4 条指令，对外部硬件而言，分别表示不同的操作，会产生 4 种不同的动作，但是编译器却会对 4 条语句进行优化，认为只有 XBYTE[2]=0x58 有效（即忽略前 3 条语句，只产生一条机器代码）。如果键入 volatile，则编译器会逐一地进行编译并产生相应的机器代码（产生 4 条代码）。

```
XBYTE[2]=0x55;
XBYTE[2]=0x56;
XBYTE[2]=0x57;
XBYTE[2]=0x58;
```

在 Java 的并发编程中，多线程共享的成员变量或静态变量，为了保证线程安全性，需要使用 synchronized 内部监视器进行同步控制，还可以使用更加轻量级的 volatile 保护。

在 JVM 1.2 之前，Java 的内存模型实现总是从主存读取变量，这是不需要进行特别的注意的。而随着 JVM 的成熟和优化，现在多线程环境下，volatile 关键字的使用变得非常重要。

在当前的 Java 内存模型下，线程可以把变量保存在本地内存（如机器的寄存器）中，而不是直接在主存（RAM）中进行读写。这就可能造成一个线程在主存中修改了一个变量的值，而另外一个线程还继续使用它在寄存器中的变量值的复制，从而出现数据不一致的情况。

要解决这个问题，只需要把该变量声明为 volatile 即可，这就指示 JVM，这个变量是不稳定的，每次使用它都要到主存中进行读取。一般情况下，多任务环境中各任务间共享的变量都应该加 volatile 修饰。

volatile 修饰的成员变量在每次被线程访问时，都强迫线程从共享内存中重读该成员变量的值。而且，当成员变量发生变化时，强迫线程将变化值回写到共享内存。这样在任何时刻，两个不同的线程总是看到某个成员变量的同一个值。

Java 语言规范中指出：为了获得最佳速度，允许线程保存共享成员变量的私有复制，而且只有当线程进入或者离开同步代码块时，才会与共享成员变量的原始值进行对比。这样当多个线程同时与某个对象交互时，就必须要让线程及时得到共享成员变量的变化。而 volatile 关键字就是提示 JVM：对于这个成员变量不能保存它的私有复制，而应直接与共享成员内存交互。

使用建议：在两个或者更多的线程访问的成员变量上使用 volatile；当要访问的变量已在 synchronized 代码块中，或者为常量时，不必使用。由于使用 volatile 屏蔽掉了 JVM 中必要的代码优化，所以其在效率上比较低，因此一定在必要时才使用此关键字。

下面将 synchronized 关键字与 volatile 进行比较：

（1）volatile 是线程同步的轻量级实现，它的性能要强于 synchronized。

（2）volatile 只能修饰变量，而 synchronized 可以修饰方法以及代码块。

（3）多线程同时访问时，volatile 不会发生阻塞。而 synchronized 会出现阻塞，因为 synchronized 使用的是排他锁，而 volatile 未使用任何锁。

（4）volatile 和 synchronized 解决的问题不一致：synchronized 解决的是在某个代码块内线程对共享资源的独占性，volatile 要解决的是共享资源在多线程间的一致性。

下面示例测试了使用 volatile 的效果，操作步骤如下。

（1）定义一个整数生成器，此为抽象类。

```
abstract class IntGenerator {
    private boolean canceled = false;
    public abstract int next();
    public void cancel() {
        canceled = true;
    }
    public boolean isCanceled() {
        return canceled;
    }
}
```

（2）定义一个整数生成器的实现类。

```
class EvenGenerator extends IntGenerator {
    private int currentEvenValue = 0;
    //期望每次调用 next()的结果都是偶数
    public int next() {
        ++currentEvenValue; //此行代码，可能被编译器优化掉
        ++currentEvenValue;
        return currentEvenValue;
    }
```

}

（3）在一个任务中，调用整数生成器，期望输出的都是偶数，一旦出现奇数就结束。

```java
class EvenChecker implements Runnable {
    private IntGenerator generator;
    public EvenChecker(IntGenerator g) {
        generator = g;
    }
    public void run() {
        while (!generator.isCanceled()) {
            int val = generator.next();
            if (val % 2 != 0) {
                System.out.println(val + "为奇数!");
                generator.cancel();      //一旦出现奇数就取消所有操作
            }
        }
    }
}
```

（4）代码测试。

```java
public static void main(String[] args) {
    System.out.println("Press Control-C 退出...");
    ExecutorService exec = Executors.newCachedThreadPool();
    EvenGenerator gen = new EvenGenerator();
    for (int i = 0; i < 10; i++) {
        //10个EvenChecker对象，共享一个整数生成器
        exec.execute(new EvenChecker(gen));
    }
    exec.shutdown();
}
```

反复运行上述程序，会频繁出现输出奇数的现象：

```
Press Control-C 退出...
1037925 为奇数!
1037927 为奇数!
1051011 为奇数!
```

（5）给 EvenGenerator 的成员变量 currentEvenValue 增加 volatile 修饰，防止其出现编译器优化。

```java
class EvenGenerator extends IntGenerator {
    private volatile int currentEvenValue = 0;
    public int next() {
        ++currentEvenValue;
```

```
        ++currentEvenValue;
        return currentEvenValue;
    }
}
```

（6）类 IntGenerator 的成员变量 canceled，增加 volatile 修饰，使多线程共享的同一变量不受寄存器等中间缓存的影响。

```
abstract class IntGenerator {
    private volatile boolean canceled = false;
    public abstract int next();
    public void cancel() {
        canceled = true;
    }
    public boolean isCanceled() {
        return canceled;
    }
}
```

运行程序，测试结果如下，仍然还出现奇数现象，为什么呢？分析代码可知，在 EvenGenerator 的 next()方法中，如果有多个线程同时进入同一个整数生成器的 next()方法，就会出现 currentEvenValue 被多个线程同时修改的现象，因此很容易出现奇数。

```
Press Control-C 退出...
25831 为奇数！
20797 为奇数！
25845 为奇数！
25843 为奇数！
25839 为奇数！
```

（7）再次修改 EvenGenerator 类，在 next()方法前增加 synchronized，反复运行程序发现，不会再出现奇数现象。在这个程序中，volatile 与 synchronized 同时使用，才能保证程序的正确运行。

```
class EvenGenerator extends IntGenerator {
    private volatile int currentEvenValue = 0;
    public synchronized int next() {
        ++currentEvenValue;
        ++currentEvenValue;
        return currentEvenValue;
    }
}
```

总结：对于并发修改共享变量，volatile 无法起到线程安全的作用，应该使用可重用锁

ReentrantLock 或 synchronized 锁住代码块。但是，对于并发读取共享变量，使用 volatile 修饰共享变量，则可以使变量值的变化马上同步给其他线程，如上例的 IntGenerator 中的 canceled 属性，它只需被某个线程改变一次即可，这时如果使用 synchronized，则不如用 volatile 简单、直观。

2. CAS 原理

CAS 是 Compare And Swap 的简称，即比较再交换。使用 CAS 可以解决多线程并发下的变量同步问题，而且 CAS 比使用传统锁机制，性能要好得多。

CAS 操作包含三个操作数——内存位置（V）、原值（A）和新值（B）。即准备把指定内存位置替换成新值前，需要校验原有的旧值，防止被其他线程修改。如果校验合格，即在运算的中间过程，指定地址的值未被修改，则替换成功，否则退出。

CAS 是一种无锁算法，CAS 实现依赖的是专有 CPU 指令集，因此它的运行效率非常高！

CAS 算法在遇到 ABA 问题时，会出现判断错误，即 CAS 在有些特例场景下是无法判断数据是否被修改的，需要小心。

3. ABA 问题

ABA 问题描述如下：

（1）若线程 1 从内存 Y 中取出 A。

（2）线程 2 也从内存 Y 中取出 A，然后线程 2 将内存 Y 中的值变更为 B，接着线程 2 又将内存 Y 中的数据修改为 A。

（3）线程 1 进行 CAS 操作发现内存 Y 中仍然是 A，然后线程 1 替换操作成功。

虽然线程 1 的 CAS 操作成功，但是整个过程是有问题的。比如，虽然链表的头在变化了两次后恢复了原值，但这并不代表链表就没有变化。所以 Java 中提供了 AtomicStampedReference、AtomicMarkableReference 来处理会发生 ABA 问题的场景，即在对象中额外再增加一个版本号来标识对象是否有过变更。

4. AtomicInteger 与 CAS

在 java.util.concurrent.atomic 包中提供了 AtomicInteger、AtomicBoolean、AtomicLong、AtomicIntegerArray 等很多用于在高并发环境下进行原子变化值的类。

以 AtomicInteger 类为例，它可以使得一个 int 值原子更新，即使有很多线程同时在修改 AtomicInteger 对象的值，也不会出现任何错误。

```java
public class AtomicInteger
        extends Number
        implements java.io.Serializable {
    public final int getAndIncrement() {}
    public final int getAndDecrement() {}
    public final int getAndAdd(int delta) {}
    public final int incrementAndGet() {}
    public final int decrementAndGet() {}
}
```

参考 JDK1.8 的 AtomicInteger 的源代码可发现，它的底层实现，依赖的都是 CAS 算法：

```java
public final int getAndIncrement() {
    return unsafe.getAndAddInt(this, valueOffset, 1); //CAS算法
}
public final int getAndDecrement() {
    return unsafe.getAndAddInt(this, valueOffset, -1);
}
```

参考 JDK1.6 的 AtomicInteger 的源代码可以发现，虽然实现方式稍有不同，但本质仍然是依赖 CAS 来实现 int 值的原子变更，如果调用 compareAndSwapInt()方法替换失败，程序会循环等待，直到重新赋值成功。

```java
public final int getAndIncrement() {
    for (;;) {
        int current = get();
        int next = current + 1;
        if (compareAndSet(current, next))
            return current;
    }
}
public final boolean compareAndSet(int expect, int update) {
    return unsafe.compareAndSwapInt(this, valueOffset, expect, update);
}
```

5. Unsafe 类源码分析

Java 内存的分配与垃圾回收依赖于 JVM，而 Java 中没有指针，因此无法直接操作内存，因此 JVM 的底层代码都是通过 C 或 C++语言实现的。

Unsafe 类通过 JNI 的方式访问本地的 C++实现库，从而使 Java 具有了直接操作内存空间的能力。但这同时也带来了一定的问题，如果不合理地使用 Unsafe 类操作内存空间，可能导致内存泄漏和指针越界，这可能导致程序崩溃，因此不推荐开发者直接调用 Unsafe。

Unsafe 部分源代码如下：

```java
import sun.misc.Unsafe;
public final class Unsafe {
    private static final Unsafe theUnsafe;
    private static native void registerNatives();
    private Unsafe() {
    }
    public static Unsafe getUnsafe() {
        Class var0 = Reflection.getCallerClass();
        if(!VM.isSystemDomainLoader(var0.getClassLoader())) {
```

```
            throw new SecurityException("Unsafe");
        } else {
            return theUnsafe;
        }
    }
}
```

注意，通过直接调用 getUnsafe() 方法，并不能获取 Unsafe 对象，原因是 VM.isSystem-DomainLoader(var0.getClassLoader()) 这句代码，会检查调用者的类加载器是否已启动。因为我们自己创建的类不是由启动加载器加载，而是默认由系统加载器加载的，所以会抛出一个异常。

```
Unsafe us = Unsafe.getUnsafe();   //会抛出异常
```

可以使用反射技术获取 Unsafe 对象，参见如下代码：

```java
public static void main(String[] args){
    try {
        Field theUnsafe = Unsafe.class.getDeclaredField("theUnsafe");
        theUnsafe.setAccessible(true);
        Unsafe us = (Unsafe) theUnsafe.get(null);
        System.out.println(us);
    } catch (Exception e) {
        e.printStackTrace();
    }
}
```

在 Unsafe 中存在着大量的方法，它们大多数是与底层打交道的，下面简单介绍几个方法：

```java
public native long allocateMemory(long var1);
//分配内存，参数为需要分配内存的字节数
public native long reallocateMemory(long var1, long var3);
//重新分配内存
public native void freeMemory(long var1);
//释放内存
public native long objectFieldOffset(Field var1);
//获取对象属性的偏移量
public void copyMemory(long var1, long var3, long var5);
//内存赋值
public native void setMemory(Object var1, long var2, long var4, byte var6);
//将给定的内存设置为固定的值
```

为了保证内存的可见性，Java 编译器在生成指令序列的适当位置会插入内存屏障指令类

禁止特定类型的处理器重排序，以下 Unsafe 方法涉及了这些内容。

```java
public native void loadFence();
//确保在屏蔽之前加载（读操作），屏蔽之后的读写操作不会重排序
public native void storeFence();
//确保在屏蔽之前加载和存储（读写操作），屏蔽之后的写操作不会重排序
public native void fullFence();
//确保在屏蔽之前读写操作，屏蔽之后的读写操作不会重排序
```

6. 案例：AtomicInteger 设置自增长 ID

创建一个 Map 集合，主键存储自增长的 ID，值为线程名称。
操作步骤如下：

（1）自定义集合 MyMap，当存入的主键冲突时，抛出异常。

```java
public class MyMap<K,V> extends HashMap<K,V>{
    @Override
    public V put(K key, V value) {
        V v = this.get(key);
        if(v != null) {
            throw new RuntimeException(key + "已存在...");
        }
        return super.put(key, value);
    }
}
```

（2）定义静态变量，所有线程获取的主键 ID 都从 sNum 获取。

```java
public static int sNum = 0;
```

（3）在主函数中，定义集合，用于存储所有生成的 ID。

```java
Map<String,Object> map = new MyMap();
```

（4）从线程池中提取线程，此处模拟 3000 个线程（线程数量少不会出现并发冲突）。多线程并发写入 ID 到 Map 集合中。

```java
ExecutorService pool = Executors.newCachedThreadPool();
for(int i=0;i<3000;i++) {
    pool.execute(new Runnable() {
        @Override
        public void run() {
            sNum++;
            String key = String.valueOf(sNum);
            map.put(key,Thread.currentThread().getName());
        }
```

```
        });
    }
pool.shutdown();
```

（5）稍作延时后，读取 Map 集合中的元素信息和元素总数。

```
try {
    Thread.sleep(500);
} catch (Exception e) {
    e.printStackTrace();
}
for (Map.Entry<String,Object> entry : map.entrySet()) {
    System.out.println(entry.getKey() + "---" + entry.getValue());
}
System.out.println("元素数量: " + map.size());
```

程序运行结果如下，由于多线程共享的变量 sNum 没有做任何同步保护，因此会出现多线程读取到相同数据的情况：

```
Exception in thread "pool-1-thread-13"
        java.lang.RuntimeException: 275 已存在...
        at com.icss.aqs.MyMap.put(MyMap.java:10)
Exception in thread "pool-1-thread-114"
        java.lang.RuntimeException: 1021 已存在...
        at com.icss.aqs.MyMap.put(MyMap.java:10)
Exception in thread "pool-1-thread-131"
        java.lang.RuntimeException: 1020 已存在...
        at com.icss.aqs.MyMap.put(MyMap.java:10)
1462---pool-1-thread-119
2793---pool-1-thread-147
1461---pool-1-thread-23
2792---pool-1-thread-147
2781---pool-1-thread-147
2780---pool-1-thread-147
2773---pool-1-thread-147
1441---pool-1-thread-11
2772---pool-1-thread-147
1449---pool-1-thread-11
...
元素数量: 2975
```

（6）为了解决步骤（5）出现的主键冲突错误，使用 synchronized 锁定同步代码块。

```
public void run() {
    synchronized (map) {
```

```
        sNum++;
        String key = String.valueOf(sNum);
        map.put(key,Thread.currentThread().getName());
    }
}
```

反复运行上面的程序，输出结果如下，不会再出现主键冲突错误。

```
1443---pool-1-thread-63
2774---pool-1-thread-86
1442---pool-1-thread-61
2773---pool-1-thread-29
1441---pool-1-thread-61
...
2772---pool-1-thread-86
1449---pool-1-thread-3
元素数量：3000
```

（7）使用 synchronized 锁定同步代码块，虽然可以解决主键冲突问题，但是使用锁会带来线程阻塞排队，这在高并发程序中需要消耗较多的性能。下面我们使用 AtomicInteger 来解决并发冲突问题。

```
public static AtomicInteger sNum = new AtomicInteger(0);
...
public void run() {
    int iKey = sNum.incrementAndGet();
    String key = String.valueOf(iKey);
    map.put(key, Thread.currentThread().getName());
}
```

反复运行上面的程序，不会再出现主键冲突的异常，但是元素数量却始终不为 3000，这是什么问题呢？

```
2774---pool-1-thread-63
1442---pool-1-thread-67
2773---pool-1-thread-62
1441---pool-1-thread-67
...
2772---pool-1-thread-62
1449---pool-1-thread-35
元素数量：2993
```

（8）检查代码发现，MyMap 的父类为 HashMap，而 HashMap 中没有并发控制代码，这会造成高并发环境向 HashMap 中插入数据时，会有数据遗漏现象（不会抛出异常信息）。

因此修改如下：

```java
public class MyMap<K,V> extends ConcurrentHashMap<K,V>{}
```

反复测试步骤（8）中的代码，使用 AtomicInteger 与 ConcurrentHashMap，在高并发向集合写入数据时，不会再出现任何错误。而且，这套代码组合，实现了性能最优！

```
2777---pool-1-thread-50
1445---pool-1-thread-26
2776---pool-1-thread-20
1444---pool-1-thread-26
2775---pool-1-thread-46
2773---pool-1-thread-30
1441---pool-1-thread-26
...
2772---pool-1-thread-56
1449---pool-1-thread-59
元素数量：3000
```

7.3.2 阻塞

在 JSR166 之前，阻塞线程和解除线程阻塞都是基于 Java Object 内置的监视器，没有基于 Java API 创建的同步器。唯一可以选择的是 Thread.suspend 和 Thread.resume，但是它们都有无法解决的竞态问题（这两个方法后期已被废除）。

JUC 包有一个 LockSuport 类，这个类中包含了解决这个问题的方法。LockSupport.park 阻塞当前线程直到有个 LockSupport.unpark 方法被调用。unpark 的调用是没有被计数的，因此在一个 park 调用前多次调用 unpark 方法只会解除一个 park 操作。在缺少一个 unpark 操作时，下一次调用 park 就会阻塞。

park 方法同样支持超时设置，以及与 JVM 的 Thread.interrupt 结合，可通过中断来 unpark 一个线程。

```java
public class LockSupport {
    private static final sun.misc.Unsafe UNSAFE;
    private LockSupport() {} //不能实例
    public static void park() {
        UNSAFE.park(false, 0L);
    }
    public static void unpark(Thread thread) {
        if (thread != null)
            UNSAFE.unpark(thread);
    }
    public static void parkUntil(long deadline) {
```

```
        UNSAFE.park(true, deadline);
    }
}
```

LockSupport 阻塞线程与解阻塞,测试步骤如下:

(1) 创建线程 t1,当 i=5 时,调用 LockSupport.park()阻塞当前线程。

```
Thread t1 = new Thread(new Runnable() {
    public void run() {
        for(int i=0;i<10;i++) {
            System.out.println(Thread.currentThread().getId()
                            + ",i=" +i);
            if(i==5) {
                System.out.println(Thread.currentThread().getId()
                                + "开始等待...");
                LockSupport.park();
            }
        }
    }
});
```

(2) 创建线程 t2,调用 LockSupport.*unpark*(t1),给 t1 线程解锁。

```
Thread t2 = new Thread(new Runnable() {
    public void run() {
        System.out.println(Thread.currentThread().getId() + " running...");
        System.out.println(Thread.currentThread().getId()+",发送解锁通知...");
        LockSupport.unpark(t1);
    }
});
```

(3) 启动线程 t1,稍作延时后,启动线程 t2。

```
t1.start();
try {
    Thread.sleep(200);
} catch (Exception e) {
}
System.out.println(t1.getId()+ "状态: " + t1.getState());
t2.start();
```

程序运行结果如下,将 LockSupport 的 park()/unpark()方法与 Object 的 wait()/notify()方法进行对比发现,功能基本一致。

```
8,i=0
```

```
8,i=1
8,i=2
8,i=3
8,i=4
8,i=5
8 开始等待...
8 状态: WAITING
9 running...
9,发送解锁通知...
8,i=6
8,i=7
8,i=8
8,i=9
```

7.3.3 排队

AQS 的核心就是如何使用队列管理被阻塞的线程，该队列是严格的 FIFO 队列，因此 AQS 不支持基于优先级的同步。

同步队列的最佳选择是自身不使用底层锁来构造非阻塞数据结构，目前，业界对此很少有争议。而其中主要有两个选择：一个是 Mellor-Crummey 和 Scott 锁（MCS 锁）的变体，另一个是 Craig、Landin 和 Hagersten 锁（CLH 锁）的变体。

一直以来，CLH 锁仅被用于自旋锁。但是，在 AQS 中，CLH 锁显然比 MCS 锁更合适。因为 CLH 锁可以更容易地去实现"取消"和"超时"功能。

CLH 队列实际上并不那么像队列，因为它的入队和出队操作都与锁紧密相关。它是一个链表队列，通过两个字段 head 和 tail 来存取，这两个字段是可原子更新的，两者在初始化时都指向一个空节点（见图 7-1）。

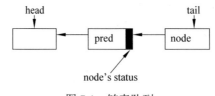

图 7-1 链表队列

一个新节点 Node，通过原子操作入队。

```
do {
    pred = tail;
} while(!tail.compareAndSet(pred, node));
```

每个节点的"释放"，状态都保存在其前驱节点中，因此，自旋锁的"自旋"操作如下：

```
while (pred.status != RELEASED);   //自旋
```

自旋等待后的出队操作，只需将 head 字段指向刚刚得到锁的节点即可：

```
head = node;
```

CLH 锁的优点在于其入队和出队操作是快速、无锁、无障碍的。即使在多线程竞争环境下，某个线程也总会赢得一次插入机会，从而能继续执行。

为了将 CLH 队列用于阻塞式同步器，需要做些额外的修改以提供一种高效的方式定位某个节点的后继节点。在自旋锁中，一个节点只需要改变其状态，下一次自旋中其后继节点就能注意到这个改变，所以节点间的链接并不是必需的。但在阻塞式同步器中，一个节点需要显式地唤醒（unpark）其后继节点。

AQS 队列的每个节点都包含一个 next 指向它的后继节点。但是，由于没有针对双向链表节点的类似 compareAndSet 的原子性无锁插入指令，因此这个 next 的设置并非作为原子性插入操作的一部分，而仅是在节点被插入后进行简单赋值：

```
pred.next = node;
```

next 仅是一种优化。如果通过某个节点的 next 字段发现其后继节点不存在，总是可以使用 pred 字段从尾部开始向前遍历来检查是否真的有后续节点。

第二个对 CLH 队列的修改是将每个节点都有的状态字段用于控制阻塞而非自旋。在 AQS 框架中，仅在线程调用具体子类中的 tryAcquire 方法返回 true 时，队列中的线程才能从 acquire 操作中返回；而单一的释放标记位是不够的，仍然需要做些控制以确保当一个活动的线程位于队列头部时，仅允许其调用 tryAcquire；这时的 acquire 可能会失败，然后阻塞。这种情况不需要读取状态标识，因为可以通过检查当前节点的前驱是否为 head 来确定权限。与自旋锁不同，CPU 读取 head 以保证复制时不会有太多的内存竞争。

用于线程排队的节点，主要源代码参考如下：

```
final class Node {
    static final Node SHARED = new Node();
    static final Node EXCLUSIVE = null;
    static final int CANCELLED =  1;
    static final int SIGNAL    = -1;
    static final int CONDITION = -2;
    static final int PROPAGATE = -3;
    volatile int waitStatus;
    volatile Node prev;
    volatile Node next;
    volatile Thread thread;
    Node nextWaiter;
    final boolean isShared() {
        return nextWaiter == SHARED;
    }
```

```java
        final Node predecessor() throws NullPointerException {
            Node p = prev;
            if (p == null)
                throw new NullPointerException();
            else
                return p;
        }
        Node(Thread thread, Node mode) {
            this.nextWaiter = mode;
            this.thread = thread;
        }
    }
```

7.3.4 条件队列

AQS 框架提供了一个条件对象（ConditionObject）类，服务于实现 Lock 接口的排他锁同步器。一个锁对象可以关联任意数目的条件对象,可以提供典型的 await、signal 和 signalAll 等操作，以及一些检测、监控的方法。

ConditionObject 有效地将条件（condition）与其他同步操作结合到了一起。只有当前线程持有锁且要操作的条件属于该锁时，条件操作才是合法的。

一个 ConditionObject 关联到一个 ReentrantLock 上，这与调用 Object.wait 构建监视器阻塞当前线程的功效一样（前面我们已经学习过了 Condition 的 await()和 signal()方法）。

ConditionObject 类是 AQS 类的成员内部类，主要方法参考如下：

```java
public class ConditionObject
            implements Condition, java.io.Serializable {
    private transient Node firstWaiter;
    private transient Node lastWaiter;
    public final void signal() { }
    public final void signalAll() {}
    public final void await()
                throws InterruptedException {}
    public final boolean awaitUntil(Date deadline)
                throws InterruptedException {}
    public final boolean await(long time, TimeUnit unit)
                throws InterruptedException {}
}
```

7.4　使用 AQS

AQS 类将上述功能和性能目标结合到一起，基于设计模式的 template method（模板方法），作为基类提供给同步器。子类只需实现控制 acquire 和 release 操作的状态检查、更新等代码。

所有 java.util.concurrent 包中的同步器类都声明了一个私有的继承了 AbstractQueuedSynchronizer 的内部类，并且把所有同步方法都委托给这个内部类来完成。这样各个同步器类的公开方法就可以使用适合自己的名称了。

参见如下 JDK 中已有的同步器类定义：

```
public class ReentrantLock
        implements Lock, java.io.Serializable {
    abstract static class Sync
            extends AbstractQueuedSynchronizer {}
    ...
}
public class Semaphore
        implements java.io.Serializable {
    abstract static class Sync
            extends AbstractQueuedSynchronizer {}
    ...
}
public class CountDownLatch {
    private static final class Sync
            extends AbstractQueuedSynchronizer {}
    ...
}
public class ReentrantReadWriteLock
        implements ReadWriteLock, java.io.Serializable {
    abstract static class Sync
            extends AbstractQueuedSynchronizer {}
    ...
}
```

下面我们自定义一个最简单的 Mutex 类的实现，它使用同步状态 0 表示解锁，1 表示锁定。这个类并不需要同步方法中的参数，因此这里在调用的时候使用 0 作为实参，方法实现里将其忽略。

```
class Mutex {
    private final Sync sync = new Sync();
```

```java
class Sync extends AbstractQueuedSynchronizer {
    public boolean tryAcquire(int ignore) {
        return compareAndSetState(0, 1);
    }
    public boolean tryRelease(int ignore) {
        setState(0);
        return true;
    }
}
public void lock() {
    sync.acquire(0);
}
public void unlock() {
    sync.release(0);
}
}
```

AbstractQueuedSynchronizer 类也提供了一些方法用来协助策略控制。例如，基础的 acquire 方法有可超时和可中断的版本。虽然到目前为止，我们的讨论都集中在像锁这样的独占模式的同步器上，但 AbstractQueuedSynchronizer 类也包含 acquireShared 等方法，它们的不同点在于 tryAcquireShared 和 tryReleaseShared 方法通过返回值，能够告知框架尚能接收多少请求，最终框架会通过级联的 signal 唤醒多个线程。

虽然将同步器序列化一般来说没有太大意义，但这些类经常会被用于构造其他类，如线程安全的集合，而这些集合通常是可序列化的。AbstractQueuedSynchronizer 和 ConditionObject 类都提供了方法用于序列化同步状态，但不会序列化被阻塞的线程，也不会序列化其他内部暂时性的变量。而且，在反序列化时，大部分同步器类也只将同步状态重置为初始值。

7.4.1 控制公平性

尽管同步器是基于 FIFO 队列的，但它们并不一定是公平的。在基础的 acquire 算法中，tryAcquire 是在入队前被执行的。因此一个新的 acquire 线程能够"窃取"本该属于队列头部第一个线程的进入同步器的机会。

可竞争抢夺的 FIFO 策略通常会提供比其他技术更高的吞吐量。当一个有竞争的锁已经空闲，而下一个准备获取锁的线程又正在解除阻塞的过程中，这时就没有线程可以获取到这个锁。如果使用抢夺策略，则可减少这之间的时间间隔。与此同时，这种策略还可避免过分的、无效率的竞争。在只要求短时间持有同步器的场景中，创建同步器的开发者可以通过定义 tryAcquire 在控制权返回之前重复调用自己若干次，来进一步凸显抢夺效果（见图 7-2）。

可抢夺的 FIFO 同步器只有概率上的公平属性。锁队列头部第一个解除了阻塞的线程拥有一次机会来赢得与闯入线程之间的竞争。如果竞争失败，要么重新阻塞、要么再次重试。

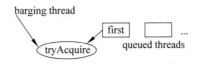

图 7-2　抢夺同步器

然而，如果闯入的线程到达的速度比队列头的线程解阻塞的速度快，那么在队列中的第一个线程将很难赢得竞争，以至于几乎总要重新阻塞，并且它的后继节点也会一直保持阻塞。对于短暂持有的同步器来说，在队列中第一个线程被解除阻塞期间，多处理器上很可能发生过多次闯入和 release 了，因此保持一个或多个线程高速运行的同时，要尽量避免其他线程饥饿的发生。

当有更高的公平性需求时，你可以把 tryAcquire 方法定义为：若当前线程不是队列的头节点，则立即失败，返回 false 即可。

一个更快，但并非严格公平的变体可以这样做，若判断瞬间队列为空，允许 tryAcquire 执行成功。在这种情况下，多个线程同时遇到一个空队列时可能会去竞争以使自己第一个获得锁，这样通常至少有一个线程是无须入队列的。JUC 包中所有支持公平模式的同步器都采用了这种策略。

7.4.2　同步器

ReentrantLock 类使用 AQS 同步状态来保存锁持有的次数。当锁被一个线程获取时，ReentrantLock 也会记录下当前获得锁的线程标识，以便检查是不是重复获取，以及当错误的线程试图进行解锁操作时检测是否存在非法状态异常。

ReentrantLock 也使用了 AQS 提供的 ConditionObject，还向外暴露了其他监控方法。ReentrantLock 通过在内部声明两个不同的 AQS 实现类来实现可选的公平模式，在创建 ReentrantLock 实例的时候可以指定是否采用公平模式。

```
public class ReentrantLock
        implements Lock, java.io.Serializable {
    abstract static class Sync
            extends AbstractQueuedSynchronizer {
    }
    static final class NonfairSync extends Sync {
    }
    static final class FairSync extends Sync {
    }
    public ReentrantLock() {
        sync = new NonfairSync();
    }
    public ReentrantLock(boolean fair) {
        sync = fair ? new FairSync() : new NonfairSync();
```

        }
    }

ReentrantReadWriteLock 类使用 AQS 同步状态 state 中的 16 位来保存写锁持有的次数，剩下的 16 位用来保存读锁的持有次数。WriteLock 的构建方式同于 ReentrantLock，ReadLock 则通过使用 acquireShared 方法来同时支持多个读线程。

Semaphore 类使用 AQS 同步状态 state 来保存信号量的当前计数。它里面定义的 acquireShared 方法会减少计数，或当计数为非正值时阻塞线程；tryRelease 方法会增加计数，可能在计数为正值时还要解除线程的阻塞。

CountDownLatch 类使用 AQS 同步状态 state 来表示计数。当该计数为 0 时，所有的 acquire 操作才能通过。

FutureTask 类使用 AQS 同步状态 state 来表示某个异步计算任务的运行状态（初始化、运行中、被取消和完成）。设置或取消一个 FutureTask 时会调用 AQS 的 release 操作，等待计算结果的线程阻塞解除是通过 AQS 的 acquire 操作实现的。

SynchronousQueues 类使用了内部等待节点，这些节点可以用于协调生产者和消费者。同时，它使用 AQS 同步状态 state 来控制当某个消费者消费时，允许一个生产者继续生产，反之亦然。

JUC 包的使用者当然可以定义自己的同步器。例如，那些曾考虑到过，但没有采纳进这个包的同步器，如提供 WIN32 事件各种风格的语义类、二元信号量、集中管理的锁以及基于树的屏障等。

7.5　AQS 性能

AQS 框架除了支持互斥锁外，还支持其他形式的同步方式，锁的性能是最容易测量和比较的，当然存在许多不同的测量方式。这里的实验主要是设计来展示锁的开销和吞吐量。

在每个测试中，所有线程都重复地更新一个伪随机数，该随机数由 nextRandom(int seed) 方法计算：

```
int t = (seed % 127773) * 16807 - (seed / 127773) * 2836;
return (t > 0)? t : t + 0x7fffffff;
```

在每次迭代中，线程以概率 S 在一个互斥锁下更新共享的生成器，否则更新其自己局部的生成器，此时是不需要锁的。如此，锁占用区域的耗时是短暂的，这就使线程持有锁期间被抢占的外界干扰降到了最小。这个函数的随机性服务于两个目的：决定是否需要使用锁，以及使循环中的代码不能被轻易地优化掉。

这里比较了四种锁：内置锁，用的是 synchronized 块；互斥锁，用的是前面讲的简单 Mutex 类；可重入锁，用的是 ReentrantLock；公平锁，用的是 ReentrantLock 的公平模式。

所有测试都运行在 JDK1.5 的服务器模式下。在收集测试数据前，测试程序先运行 20

次非竞争的测试,以排除 JVM "预热"过程的影响。除了公平模式下的测试只跑了一百万次迭代外,其他每个线程中的测试都运行了一千万次迭代。

该测试运行在四个 X86 机器和四个 UltraSparc 机器上。所有 X86 机器都运行的是 RedHat 基于 NPTL 2.4 内核和库的 Linux 系统。所有的 UltraSparc 机器都运行的是 Solaris-9。测试时所有系统的负载都很轻。表 7-1 中 "4p" 这个名字反映出双核超线程的 Xeon 更像是 4 路机器,而不是 2 路机器。这里没有将测试数据规范化,见表 7-1,同步的相对开销与处理器的数量、类型、速度之间不具备简单的关系。

表 7-1 测试平台环境

机 器 名	处 理 器	处理器类型	速度/MHz
1p	1	Pentium3	900
2p	2	Pentium3	1400
2A	2	Athlon	2000
4p	2HT	Pentium4/Xeon	2400
1U	1	UltraSparc2	650
4U	4	UltraSparc2	450
8U	8	UltraSparc3	750
24U	24	UltraSparc3	750

7.5.1 过载

无竞争情况下的性能开销是仅运行一个线程,将概率 S 为 1 时的每次迭代时间减去概率 S 为 0 时的每次迭代时间即可(概率为 0 时是没有锁操作,概率为 1 时是每次都有锁操作)。表 7-2 以 ns 为单位,显示了非竞争场景下每次锁操作的开销。Mutex 类最接近于框架的基本耗时,可重入锁的额外开销是记录当前所有者线程和错误检查的耗时,对于公平锁来说还包含检查队列是否为空的耗时。

表 7-2 无竞争时单锁开销   单位:ns

机 器	内置锁耗时	互斥锁耗时	可重入锁耗时	公平可重入锁耗时
1p	18	9	31	37
2p	58	71	77	81
2A	13	21	31	30
4p	116	95	109	117
1U	90	40	58	67
4U	122	82	100	115
8U	160	83	103	123
24U	161	84	108	119

表 7-2 也展示了内置锁 tryAcquire 的耗时。这里的差异主要反映出了不同锁和不同机器使用的不同的原子指令以及内存屏障的耗时。在多处理器上，这些指令常常是完全优于所有其他指令的。内置锁和同步器类之间的主要差别，显然是由于 Hotspot 锁在锁定和解锁时都使用了一次 compareAndSet，而同步器的 acquire 操作使用了一次 compareAndSet，但 release 操作用的是一次 volatile。每个锁的绝对和相对耗时因机器的不同而不同。

表 7-3 展示了概率 S 为 1，运行 256 个并发线程时产生的大规模锁竞争下每个锁的开销。在完全饱和的情况下，可抢夺的 FIFO 锁比内置锁的开销少了一个数量级（也就是有更大的吞吐量），比公平锁更是少了两个数量级。这表现了当存在极大的并发竞争时，可抢夺 FIFO 策略有很大的优势。

表 7-3　饱和时单锁开销　　　　　　　　　　　　　　　　　　　　单位：ns

机　　器	内置锁耗时	互斥锁耗时	可重入锁耗时	公平可重入锁耗时
1p	521	46	67	8327
2p	930	108	132	14967
2A	748	79	84	33910
4p	1146	188	247	15328
1U	879	153	177	41394
4U	2590	347	368	30004
8U	1274	157	174	31084
24U	1983	160	182	32291

表 7-3 也说明了即使在内部开销比较低的情况下，公平锁的性能也完全是由上下文切换的时间所决定的。列出的时间大致上都与各平台上线程阻塞和解阻塞的时间对应。

此外，后面增加的一个实验（仅使用机器 4p）表明，对于这里用到的短暂持有的锁，公平参数的设置在总差异中的影响很小。这里将线程终止时间的差异记录成一个粗粒度的离散量数。在 4p 的机器上，公平锁的时间度量的标准差平均为 0.7%，可重入锁平均为 6.0%。作为对比，为模拟一个长时间持有锁的场景，测试中使每个线程在持有锁的情况下计算了 16000 次随机数。这时，总运行时间几乎是相同的（公平锁：9.79s，可重入锁：9.72s）。公平模式下的差异依然很小，标准差平均为 0.1%，而可重入锁上升到了标准差平均为 29.5%。

7.5.2　吞吐量

大部分同步器应用介于无竞争和极大线程竞争环境之间。这可以用实验在两个维度进行检查：修改竞争线程集合的数量，或向线程集合里增加更多的线程。为了说明这些影响，测试运行不同竞争线程数目产生的影响（都用可重入锁），参见下面的 slowdown 度量标准公式：

$$\text{slowdown} = \frac{t}{S \cdot b \cdot n + (1-S) \cdot b \cdot \max\left(1, \frac{n}{p}\right)}$$

这里，t 是总运行时间，b 是一个线程在没有竞争或同步下的基线时间，n 是线程数，p 是处理器数，S 是共享访问的比例。计算结果是实际执行时间与理想执行时间的比率，理想执行时间是通过使用 Amdahl 法则计算出来的。理想时间是模拟了一次：没有同步开销，没有因锁争用而导致线程阻塞的执行过程。即使这样，在很低的竞争下，相比理想时间，有一些测试结果却表现出了很小的速度增长，大概是由于基线和测试之间的优化、流水线等方面有着轻微的差别。

公式中用以 2 为底的对数为比例进行了缩放。例如，值为 1 表示实际时间是理想时间的 2 倍，4 表示慢 16 倍。使用对数就不需要依赖一个随意的基线时间，因此，基于不同底数计算的结果表现出的趋势应该是类似的。这些测试使用的竞争概率从 1/128 到 1，以 2 的幂为步长，线程的数量从 1 到 1024，以 2 的幂的一半为步长。

在单处理器（1p 和 1U）上，性能随着竞争的上升而下降，但不会随着线程数的增加而下降。多处理器在遭遇竞争时，性能下降得更快。根据多处理器相关测试显示，开始出现的峰值处虽然只有几个线程的竞争，但相对性能通常却最差。这反映出了一个性能的过渡区域，在这里闯入的线程和被唤醒的线程都准备获取锁，这会让它们频繁地迫使对方阻塞。在大多数时候，过渡区域后面会紧接着一个平滑区域，因为此时几乎没有空闲的锁，所以会与单处理器上顺序执行的模式差不多；在多处理器机器上会较早进入平滑区域。例如，请注意在满竞争下，这些值在处理器越少的机器上，会有更糟糕的相对速度下降。

根据这些结果，可以针对阻塞（park/unpark）做进一步调优以减少上下文切换和相关的开销，这会给 AQS 框架带来显著的性能提升。此外，在多处理器上为短时间持有的但高竞争的锁采用某种形式的适应性自旋，可以避免这里看到的一些波动，这对同步器类大有裨益。虽然在跨越不同上下文时，适应性自旋很难表现良好，但可以使用 AQS 框架为特定应用构建一个自定义形式的同步锁。

7.6 本章习题

（1）AQS 框架提供了一套同步管理的通用机制，如下哪个选项不属于 AQS？（   ）

A. 同步状态原子性管理　　　　　　B. 线程阻塞与解除阻塞
C. 线程创建管理　　　　　　　　　D. 线程排队管理

（2）如下哪个选项，不属于 AQS 的实现类？（   ）

A. ReentrantLock　　　　　　　　B. ReentrantReadWriteLock
C. Semaphore　　　　　　　　　　D. BlockingQueue
E. CyclicBarrier

（3）关于 AQS 同步状态的描述，不正确的是（   ）。

A. AQS 类使用一个 int（32 位）值来保存同步状态
B. CAS 是一种无锁算法，CAS 实现依赖的是专有 CPU 指令集
C. AQS 同步状态的读取与更新都依赖于 java.util.concurrent.atomic 包支持
D. AQS 同步状态更新需要使用 synchronized 锁保证线程安全

（4）AQS 框架，线程阻塞采用的方案是（　　）。

A. Object 对象的 wait()   B. Condition 对象的 await()
C. LockSuport 对象的 park()   D. synchronized 的隐式锁

（5）AQS 框架，关于线程排队描述错误的是（　　）。

A. AQS 使用阻塞队列进行线程排队
B. AQS 不支持基于优先级的同步
C. AQS 队列可以不满足 FIFO 特性
D. CLH 队列的入队与出队操作都与锁紧密相关

第 8 章 结束线程与线程池任务

创建任务、提交任务、启动线程、CPU 运行任务，最后是任务执行完毕、释放相关资源，这是多线程运行的标准流程。但是，有些任务在正常结束前，可能需要取消任务、结束线程，如何操作呢？

在 Thread 类中，原有的一些相关方法如 destroy()、stop()等已被废弃。如何安全、可靠、快速地取消任务、停止线程并不容易。JDK 以前的线程停止机制存在严重问题，因此被迫作废。在当前的 JDK 最新版本中，仍然没有非常有效的手段，使线程直接从 RUNNABLE、WAITING、BLOCKED 等状态直接进入 TERMINATED 状态。

```
public class Thread implements Runnable {
    @Deprecated
    public void destroy() {
        throw new NoSuchMethodError();
    }
    @Deprecated
    public final synchronized void stop(Throwable obj) {
        throw new UnsupportedOperationException();
    }
}
```

8.1 stop()与 destroy()

JDK1.5 以前版本，在 Thread 类中提供了 stop()和 destroy()方法，用于线程的停止和销毁。

stop()方法是不安全的，使用这个方法停止线程，可以解锁所有和该线程相关的监视器。那些原来受监视器保护的对象会处于不一致的状态，因此受损对象可能变得对其他线程可见，从而导致任意行为的发生。

destroy()方法的设计初衷是为了销毁线程，该线程所持有的任何监视器仍然保持锁定。但是，该方法从未真正实现。如果目标线程持有保护关键系统资源的锁，当线程被销毁后，则无法再次访问该资源，这将直接导致死锁。

8.2 状态值结束线程

参见 3.3.3 节，使用状态值结束线程运行。

8.3 shutdown()与 shutdownNow()

如何使用 shutdown()与 shutdownNow()方法关闭线程池中的任务，参见 4.1.2 节和 4.3.2 节相关内容。

8.4 线程休眠

线程休眠就是使正在执行任务的线程，进入 WAITING 状态。这时会释放当前线程占用的监视器锁，而且 CPU 不会再给该任务分配资源。线程的 WAITING 与 BLOCKED 状态完全不同，这是一种非常经济的、不占用系统资源的休眠状态。

参见 3.1 节，通过 Object 对象的 wait()/notify()方法，休眠或唤醒线程。

参见 5.2 节，通过 Condition 对象的 await()/signal()方法，休眠或唤醒线程。

8.5 线程中断

参见 3.3 节线程中断内容。

8.6 Future 与 FutureTask

参见 4.3.1 节，任务实现 Callable 接口，通过 ExecutorService 的 submit()提交任务，可以使用 Future 接收 call()方法的返回值。

```
ExecutorService pool = Executors.newCachedThreadPool();
Future<String> f = pool.submit(new Callable(){...});
String result = f.get();
```

Future 类表示异步计算的未来结果,这个结果最终将在处理完成后出现在 Future 中。返回结果只能在异步执行完成后使用 get()方法进行检索。

get()方法是阻塞等待模式，如果异步线程未执行完成相应的任务，则接收线程只能阻塞等待，直到结果返回。

```
public interface Future<V> {
    boolean cancel(boolean mayInterruptIfRunning);
```

```
    boolean isCancelled();
    boolean isDone();
    V get() throws InterruptedException, ExecutionException;
    V get(long timeout, TimeUnit unit)
        throws InterruptedException,
        ExecutionException, TimeoutException;
}
```

线程池 ThreadPoolExecutor 的 submit()方法，创建的是 FutureTask 实例，FutureTask 类实现了 Future 接口，FutureTask 的部分源码实现如下。

```
public class FutureTask<V> implements RunnableFuture<V> {
    /*状态值 state 变化参考：
    * NEW -> COMPLETING -> NORMAL
    * NEW -> COMPLETING -> EXCEPTIONAL
    * NEW -> CANCELLED
    * NEW -> INTERRUPTING -> INTERRUPTED*/
    private volatile int state;
    public V get() throws InterruptedException, ExecutionException {
        int s = state;
        if (s <= COMPLETING)
            s = awaitDone(false, 0L);     //阻塞等待
        return report(s);
    }
    public boolean cancel(boolean mayInterruptIfRunning) {
        if (mayInterruptIfRunning) {
            try {
                Thread t = runner;
                if (t != null)
                    t.interrupt();        //通过中断尝试结束线程
            } finally {
                UNSAFE.putOrderedInt(this, stateOffset, INTERRUPTED);
            }
        }
        return true;
    }
}
```

8.6.1 取消任务

调用 Future 中的 cancel()方法，可以取消正在执行中的异步任务。
代码测试步骤如下。
（1）定义静态变量，用于输出执行任务的异步线程信息。

```
static Thread tmpThread;
```

（2）新建任务类 MyTask，实现 Callable 接口。

```
class MyTask implements Callable<String> {
    public String call() throws Exception {
        System.out.println("线程" + Thread.currentThread().getId() + "开始计算...");
        tmpThread = Thread.currentThread();
        int rand = (int)(Math.random()*10);
            TimeUnit.SECONDS.sleep(rand);
            String result = "线程" + Thread.currentThread().getId()
                        + "计算结果：" + Math.PI*rand;
            System.out.println("output: " + result);
            return result;
    }
}
```

（3）调用 submit()方法提交任务。延迟 3 秒后，任务如果未完成，则调用 cancel()方法取消正在执行的任务。

```
public static void main(String[] args){
    ExecutorService pool = Executors.newFixedThreadPool(10);
    for(int i=0;i<10;i++) {
        Future<String> future = pool.submit(new MyTask());
        try {
            Thread.sleep(3000);
            if(future.isDone()) {
                System.out.println("get:" + future.get());
            }else {
                future.cancel(true);
                if(future.isCancelled()) {
                    System.out.println("等待超时，任务取消！"
                            + tmpThread.getId() + ":" + tmpThread.getState());
                }
            }
            //输出任务执行完成或取消后的线程状态
            Thread.sleep(100);
            System.out.println(tmpThread.getId()+":"+tmpThread.getState());
        } catch (Exception e){
             e.printStackTrace();
        }
    }
    pool.shutdown();
```

}

程序运行结果如下，通过分析可知：任务取消时，执行任务的线程处于 TIMED_WAITING 状态；任务执行完毕或取消任务后，执行任务的线程回到线程池，处于 WAITING 待命状态；等待超时，主线程发出 cancel()指令后，异步任务都被立即结束了！

```
线程 8 开始计算...
等待超时,任务取消！ 8:TIMED_WAITING
8:WAITING
线程 9 开始计算...
output：线程 9 计算结果：3.141592653589793
get:线程 9 计算结果：3.141592653589793
9:WAITING
线程 10 开始计算...
等待超时,任务取消！ 10:RUNNABLE
10:WAITING
线程 11 开始计算...
等待超时,任务取消！ 11:TIMED_WAITING
11:WAITING
线程 12 开始计算...
output：线程 12 计算结果：0.0
get:线程 12 计算结果：0.0
12:WAITING
线程 13 开始计算...
等待超时,任务取消！ 13:TIMED_WAITING
13:WAITING
线程 14 开始计算...
等待超时,任务取消！ 14:TIMED_WAITING
14:WAITING
线程 15 开始计算...
等待超时,任务取消！ 15:TIMED_WAITING
15:WAITING
线程 16 开始计算...
等待超时,任务取消！ 16:TIMED_WAITING
16:WAITING
线程 17 开始计算...
等待超时,任务取消！ 17:TIMED_WAITING
17:WAITING
```

（4）修改步骤（2）中的 MyTask 实现，使用大型浮点计算替换线程 sleep()。

```
class MyTask implements Callable<String> {
    public String call() throws Exception {
        System.out.println("线程"+Thread.currentThread().getId()+"开始计算...");
```

```java
            tmpThread = Thread.currentThread();
            int rand = (int)(Math.random()*10);
            double d = 0;
            long all = rand*1000*1000*1000*1000*1000;
            for(long i=0;i<all;i++) {
                d += (Math.PI + Math.E)/(double)i;
            }
            String result = "线程" + Thread.currentThread().getId()
                            + "计算结果：" + Math.PI*rand;
            System.out.println("output: " + result);
            return result;
        }
    }
```

程序运行结果如下，通过分析可知：任务取消时，执行任务的线程处于 RUNNABLE 状态；任务执行完毕，线程处于 WAITING 待命状态；等待超时，主线程发出 cancel()指令后，异步线程的状态仍然为 RUNNABLE（参见线程 10、11、15），等待一会后，线程 10、11 和 15 才陆续结束并在线程内部输出了运算结果。

```
线程 8 开始计算...
output：线程 8 计算结果：12.566370614359172
get:线程 8 计算结果：12.566370614359172
8:WAITING
线程 9 开始计算...
等待超时，任务取消！9:RUNNABLE
9:RUNNABLE
线程 10 开始计算...
等待超时，任务取消！10:RUNNABLE
10:RUNNABLE
线程 11 开始计算...
等待超时，任务取消！11:RUNNABLE
11:RUNNABLE
线程 12 开始计算...
output：线程 12 计算结果：28.274333882308138
get:线程 12 计算结果：28.274333882308138
12:WAITING
线程 13 开始计算...
output：线程 13 计算结果：18.84955592153876
get:线程 13 计算结果：18.84955592153876
13:WAITING
线程 14 开始计算...
output：线程 14 计算结果：28.274333882308138
get:线程 14 计算结果：28.274333882308138
```

```
14:WAITING
线程 15 开始计算...
等待超时,任务取消! 15:RUNNABLE
output:线程 9 计算结果:25.132741228718345
15:RUNNABLE
线程 16 开始计算...
output:线程 16 计算结果:12.566370614359172
get:线程 16 计算结果:12.566370614359172
16:WAITING
线程 17 开始计算...
output:线程 17 计算结果:12.566370614359172
get:线程 17 计算结果:12.566370614359172
17:WAITING
output:线程 15 计算结果:25.132741228718345
output:线程 10 计算结果:6.283185307179586
output:线程 11 计算结果:6.283185307179586
```

总结:反复测试上述程序,并结合 cancel()方法的源代码,可以得到如下结论:调用 cancel() 方法,其实是主线程向异步线程发出了线程中断请求,只有当异步线程处于 WAITING、TIMED_WAITING 或 BLOCKED 状态时,才会响应中断请求。如果异步线程处于 RUNNABLE 状态,不会响应中断请求。

8.6.2 任务超时结束

Future 的 get(long timeout, TimeUnit unit)方法会在任务执行时间超过预期时,抛出 TimeoutException 异常,强制结束任务返回结果的接收,但是任务本身仍然继续运行。

代码测试步骤如下。

(1)创建任务类 MyTask,实现 Callable 接口。

```java
class MyTask implements Callable<String> {
    public String call() throws Exception {
        System.out.println("线程"+Thread.currentThread().getId()+"开始计算...");
        int rand = (int)(Math.random()*10);
        TimeUnit.SECONDS.sleep(rand);
        String result = "线程" + Thread.currentThread().getId()
                     + "计算结果:" + Math.PI*rand;
        return result;
    }
}
```

(2)创建线程池,用 submit 提交异步任务,Future 接收异步返回信息。

```java
public static void main(String[] args){
```

```java
        ExecutorService pool = Executors.newFixedThreadPool(10);
        for(int i=0;i<10;i++) {
            Future<String> future = pool.submit(new MyTask());
            try {
                String result = future.get(5, TimeUnit.SECONDS);
                System.out.println(result);
            }catch(TimeoutException e) {
                System.out.println("超时异常!");
            } catch(Exception e){
                e.printStackTrace();
            }
        }
        pool.shutdown();
}
```

程序运行结果如下:

线程 8 开始计算...
线程 8 计算结果: 12.566370614359172
线程 9 开始计算...
线程 9 计算结果: 0.0
线程 10 开始计算...
线程 10 计算结果: 6.283185307179586
线程 11 开始计算...
线程 11 计算结果: 12.566370614359172
线程 12 开始计算...
超时异常!
线程 13 开始计算...
线程 13 计算结果: 9.42477796076938
线程 14 开始计算...
线程 14 计算结果: 9.42477796076938
线程 15 开始计算...
超时异常!
线程 16 开始计算...
超时异常!
线程 17 开始计算...
超时异常!

8.7 项目案例:所有线程池任务暂停与重启

案例场景描述:为了充分利用公司服务器,在服务器不忙的时候,启动定时任务进行网络爬虫的数据抓取(如凌晨 2 点到 5 点),当服务器较忙时,必须暂停所有爬虫的任务。

操作步骤如下：
（1）定义一个静态变量，用于存储从线程池启动的线程和锁对象。

static Map<Thread, Object> *ts* = **new** HashMap();

（2）定义一个全局布尔值，用于控制所有任务是否启动。

static volatile boolean *isWait* = **false**;

（3）定义方法，启动所有爬虫任务。

```
public static void startTask() {
    System.out.println("所有任务运行中...");
    ExecutorService pool = Executors.newFixedThreadPool(5);
    for (int i = 0; i < 5; i++) {
        Object obj = new Object();
        pool.execute(new Runnable() {
            public void run() {
                double d = 0;
                for (long i = 0; i < Long.MAX_VALUE; i++) {
                    if (isWait) {
                        synchronized(obj) {
                            try {
                                System.out.println(Thread.currentThread()
                                            .getId() + "休眠, i=" + i);
                                ts.put(Thread.currentThread(), obj);
                                obj.wait();
                                System.out.println(Thread.currentThread()
                                            .getId() + "重启, i=" + i);
                            } catch (InterruptedException e) {
                            }
                        }
                    }
                    d += (Math.PI + Math.E) / (double) i;
                }
            }
        });
    }
}
```

（4）当服务器繁忙时，暂停爬虫任务。

```
public static void suspendTask() {
    System.out.println("暂停所有任务执行...");
    isWait = true;
    outputState();
```

```java
    }
    private static void outputState() {
        try {
            Thread.sleep(100);
        } catch (Exception e) {
        }
        for (Map.Entry<Thread, Object> entry : ts.entrySet()) {
            System.out.println(entry.getKey().getId()
                    + "状态: " + entry.getKey().getState());
        }
    }
```

(5) 当服务器空闲时，唤醒原来的爬虫任务。

```java
public static void reStartTask() {
    System.out.println("重启所有任务...");
    isWait = false;
    for (Map.Entry<Thread, Object> entry : ts.entrySet()) {
        Object obj = entry.getValue();
        synchronized(obj) {
            obj.notify();
        }
    }
    outputState();
}
```

(6) 主线程启动任务后，不断重复暂停任务、重启任务。

```java
public static void main(String[] args) {
    startTask();
    while (true) {
        try {
            Thread.sleep(2000);
        } catch (Exception e) {
        }
        suspendTask();
        try {
            Thread.sleep(2000);
        } catch (Exception e) {
        }
        reStartTask();
    }
}
```

程序运行结果如下：

```
所有任务运行中...
暂停所有任务执行...
10 休眠, i=54437724
12 休眠, i=39414233
8 休眠, i=38646244
11 休眠, i=21282564
9 休眠, i=43065940
11 状态: WAITING
9 状态: WAITING
12 状态: WAITING
8 状态: WAITING
10 状态: WAITING
重启所有任务...
11 重启, i=21282564
9 重启, i=43065940
12 重启, i=39414233
10 重启, i=54437724
8 重启, i=38646244
11 状态: RUNNABLE
9 状态: RUNNABLE
12 状态: RUNNABLE
8 状态: RUNNABLE
10 状态: RUNNABLE
...
```

8.8 本章习题

（1）关于线程结束，如下哪个选项的描述正确？（　　）
A. 调用 Thread 类的 stop()方法结束线程
B. 调用 Thread 类的 destroy()方法结束线程
C. 通过设置状态值，结束 run()方法运行，从而结束线程
D. 调用 Object 的 wait()方法结束线程

（2）结束线程池中的任务，哪个选项不合理？（　　）
A. 调用线程池的 shutdown()方法　　B. 调用线程池的 shutdownNow()方法
C. 调用 Future 接口中的 cancel()方法　　D. 通过中断结束线程池中的任务

第 9 章 Tomcat 线程池技术

Tomcat 是 Apache 组织下的顶级开源项目，是目前使用最为广泛的 Web 服务器，它具有部署简单、响应速度快、并发负载量大等优秀特性。Tomcat 是一个高并发的环境,基于 JavaEE 的 Web 项目可以部署到 Tomcat 上；同时 Tomcat 又是 JSP 和 Servlet 的容器，JSP 和 Servlet 作为重要的 HTTP 资源被部署在 Tomcat 服务器上。

为了处理访问 Web 服务器的高并发 HTTP 请求，就必须要使用线程池技术。下面我们基于 Tomcat9.0 来分析一下 Tomcat 的线程池技术。

9.1 自定义 ThreadPoolExecutor

Tomcat 自定义了线程池，它继承了 java.util.concurrent.ThreadPoolExecutor。这里新增了一个成员变量 submittedCount，它用于监控已经提交但尚未完成的任务数量，这包括已经在队列中的任务和已经交给工作线程但还未开始执行的任务，这个数字总是大于或等于 getActiveCount()的数量。

```java
public class ThreadPoolExecutor
            extends java.util.concurrent.ThreadPoolExecutor {
    private final AtomicInteger submittedCount = new AtomicInteger(0);
    public int getSubmittedCount() {
        return submittedCount.get();
    }
    public ThreadPoolExecutor(int corePoolSize, int maximumPoolSize,
      long keepAliveTime, TimeUnit unit, BlockingQueue<Runnable> workQueue){
        super(corePoolSize, maximumPoolSize, keepAliveTime, unit,
            workQueue, new RejectHandler());
        prestartAllCoreThreads();
    }
}
```

在线程池的构造函数中，创建线程池时会调用 prestartAllCoreThreads()方法，在没有任何任务的情况下，会提前启动所有核心线程。

java.util.concurrent.ThreadPoolExecutor 中的方法定义如下：

```java
public class ThreadPoolExecutor {
    public int prestartAllCoreThreads() {
        int n = 0;
        while (addWorker(null, true))
            ++n;
        return n;
    }
}
```

在自定义线程池 ThreadPoolExecutor 中，重写了 execute() 方法。这里首先对提交执行的任务进行了 submittedCount 加一。传统的 JUC 包中的 ThreadPoolExecutor，当已经启动的工作线程数达到 maximumPoolSize 数量后，继续新增任务就会抛出 RejectedExecutionException 异常。在自定义 ThreadPoolExecutor 中，当线程池抛出 RejectedExecutionException 异常后，会调用 force() 方法再次向 TaskQueue 中进行添加任务的尝试。如果添加失败，则 submittedCount 减一。

```java
public class ThreadPoolExecutor
            extends java.util.concurrent.ThreadPoolExecutor {
    public void execute(Runnable command) {
        execute(command,0,TimeUnit.MILLISECONDS);
    }
    public void execute(Runnable command, long timeout, TimeUnit unit) {
        submittedCount.incrementAndGet();
        try {
            super.execute(command);
        } catch (RejectedExecutionException rx) {
            if (super.getQueue() instanceof TaskQueue) {
                final TaskQueue queue = (TaskQueue)super.getQueue();
                try {
                    if (!queue.force(command, timeout, unit)) {
                        submittedCount.decrementAndGet();
                        throw new RejectedExecutionException(
                            sm.getString("threadPoolExecutor.queueFull"));
                    }
                } catch (InterruptedException x) {
                    submittedCount.decrementAndGet();
                    throw new RejectedExecutionException(x);
                }
            } else {
                submittedCount.decrementAndGet();
                throw rx;
```

            }
        } } }

9.2 Tomcat 任务队列

在 Tomcat 中重新定义了一个阻塞队列 TaskQueue，它继承了 LinkedBlockingQueue。参考部分代码如下：

```
public class TaskQueue extends LinkedBlockingQueue<Runnable>{
    private transient volatile ThreadPoolExecutor parent = null;
    public boolean force(Runnable o, long timeout, TimeUnit unit)
                    throws InterruptedException {
        if (parent == null || parent.isShutdown())
            throw new RejectedExecutionException(
                    sm.getString("taskQueue.notRunning"));
        return super.offer(o,timeout,unit);
    }
}
```

阻塞队列 TaskQueue 中的成员变量 parent 为自定义的线程池 ThreadPoolExecutor 对象，它的 force()方法会调用 LinkedBlockingQueue 的 offer 方法向 TaskQueue 阻塞队列中强制插入任务。

LinkedBlockingQueue 是无界阻塞队列，底层实现为链表。它的容量 capacity 是根据业务需求，人为设置的队列容量限制。

```
public class LinkedBlockingQueue<E> extends AbstractQueue<E>
            implements BlockingQueue<E>, java.io.Serializable {
    public LinkedBlockingQueue() {
        this(Integer.MAX_VALUE);
    }
    public LinkedBlockingQueue(int capacity) {
        this.capacity = capacity;
        last = head = new Node<E>(null);
    }
}
```

9.3 Tomcat 任务线程

在 Tomcat 中重新定义了线程 Thread 的子类 TaskThread，类中记录了每个 TaskThread 的创建时间（creationTime）；另外，通过静态内部类 WrappingRunnable 对任务 Runnable 进行

了包装，这个内部类的主要作用是对 StopPooledThreadException 异常进行了处理。

```java
public class TaskThread extends Thread {
 private final long creationTime;
 public final long getCreationTime() {
       return creationTime;
 }
 public TaskThread(ThreadGroup group, Runnable target, String name) {
       super(group, new WrappingRunnable(target), name);
       this.creationTime = System.currentTimeMillis();
 }
 private static class WrappingRunnable implements Runnable {
    private Runnable wrappedRunnable;
    WrappingRunnable(Runnable wrappedRunnable) {
        this.wrappedRunnable = wrappedRunnable;
    }
    @Override
    public void run() {
      try {
          wrappedRunnable.run();
      } catch(StopPooledThreadException exc) {
          log.debug("Thread exiting on purpose", exc);
      }
    }
  }
}
```

9.4　Tomcat 任务线程工厂

在 4.5 节，我们学习了 ThreadFactory 接口和默认线程工厂的实现。在 Tomcat 中，定义了一个新的任务线程工厂 TaskThreadFactory。TaskThreadFactory 实现了 ThreadFactory 接口，重写了 newThread 方法。在 newThread 中，所有新建的线程均为 TaskThread。

```java
public class TaskThreadFactory implements ThreadFactory {
    private final ThreadGroup group;
    private final AtomicInteger threadNumber = new AtomicInteger(1);
    private final String namePrefix;
    private final boolean daemon;
    private final int threadPriority;
    public TaskThreadFactory(String namePrefix, boolean daemon, int priority) {
        SecurityManager s = System.getSecurityManager();
        group = (s != null) ? s.getThreadGroup()
```

```
                            : Thread.currentThread().getThreadGroup();
        this.namePrefix = namePrefix;
        this.daemon = daemon;
        this.threadPriority = priority;
    }
    @Override
    public Thread newThread(Runnable r) {
        TaskThread t = new TaskThread(group, r, namePrefix
                        + threadNumber.getAndIncrement());
        t.setDaemon(daemon);
        t.setPriority(threadPriority);
        ...
        return t;
    }
}
```

9.5 Tomcat 连接器与线程池

连接器 Connector 是 Tomcat 的重要管理组件，它的主要功能是接收客户端的并发请求，然后由 Web 服务器分配线程让引擎 Engine（也就是 Servlet 容器）来处理这个请求，并把服务器创建的 Request 和 Response 对象传递给 Engine。Engine 处理完请求后，也会通过 Connector 将响应返回给客户端。Tomcat 连接器与引擎的工作模式，如图 9-1 所示。

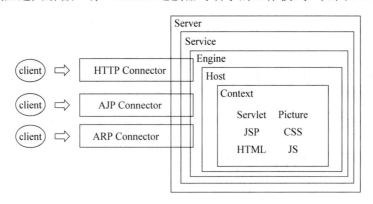

图 9-1 Tomcat 连接器

根据客户端与服务器的交互协议不同，Connector 可以分为 HTTP Connector、AJP Connector、ARP Connector 等。Connector 使用哪种 protocol，可以打开 Tomcat 的 server.xml 文件，然后修改<connector>元素的 protocol 属性进行配置，也可以使用默认值。参见如下示例：

```
<Connector connectionTimeout="20000" port="8080"
```

```
                     protocol="HTTP/1.1"  redirectPort="8443"/>
<Connector port="8443" protocol="org.apache.coyote.http11.Http11NioProtocol"
                      maxThreads="150" SSLEnabled="true"/>
<Connector port="8443" protocol="org.apache.coyote.http11.Http11AprProtocol"
                      maxThreads="150" SSLEnabled="true" >
     <UpgradeProtocol className="org.apache.coyote.http2.Http2Protocol" />
</Connector>
<Connector port="8009" protocol="AJP/1.3" redirectPort="8443"/>
```

连接器需要处理大量的并发请求，因此需要使用线程池，通过 Executor 可以定义一个或多个命名线程池。Executor 元素代表 Tomcat 中的线程池，可以由其他组件共享使用。连接器要使用线程池，需要通过 executor 属性指定它的名字。Executor 是 Service 元素的内嵌元素。为了使 Connector 能正常使用线程池，Executor 元素应该放在 Connector 前面。

Executor 与 Connector 的配置示例（未给出全部参数）如下：

```
<Executor name="tomcatThreadPool" namePrefix="catalina-exec-"
                      maxThreads="150" minSpareThreads="4"/>
<Connector executor="tomcatThreadPool"  port="8080" protocol="HTTP/1.1"
       connectionTimeout="20000" redirectPort="8443" acceptCount="1000" />
```

Connector 主要参数配置信息如下。

- port：端口号，Tomcat 默认端口号是 8080。
- address：配置 Connector 监听哪个服务器地址。服务器的 IP 可能不止一个，如果不配置 address 的话，默认会在所有的 IP 上监听请求。
- protocol：协议，默认使用 HTTP1.1。
- connectionTimeOut：客户端连接超时时间，单位是毫秒。
- acceptCount：当并发请求量大时，没有空闲线程，请求就要排队了。acceptCount 是用来配置排队的长度的。如果排队的请求数超过了配置项，其余的请求会被拒绝，这时 Connector 就不会处理新的请求了。
- maxConnections：是指 Connector 同时支持的最大连接数。对 BIO Connector 来说是线程池的最大值。当超过这个数值时，只会被接收不会被处理，必须等前面请求处理完了，有空闲线程了才会去处理。maxConnections=-1 表示不限制最大连接数，但具体的连接数还会受限于其他资源。

Executor 的主要属性信息如下。

- name：该线程池的标记。
- maxThreads：线程池中最大活跃线程数，默认值为 150。
- minSpareThreads：线程池中保持的最小线程数。
- maxIdleTime：线程空闲的最大时间，当空闲超过该值时关闭线程（除非线程数小于 minSpareThreads），单位是 ms。

- daemon：是否后台线程，默认值为 true。
- threadPriority：线程优先级，默认值为 5。
- namePrefix：线程名字的前缀，线程池中线程名字为：namePrefix+线程编号。

Tomcat 连接器的源代码核心实现如下：

```java
public class Connector extends LifecycleMBeanBase {
    protected Service service = null;              //连接器所属服务
    protected long asyncTimeout = 30000;           //默认异步请求的倒计时时间
    protected String proxyName = null;
    protected int proxyPort = 0;
    private int maxCookieCount = 200;              //HTTP 请求中 Cookie 的最大数量
    protected int maxParameterCount = 10000;       //HTTP 请求参数的最大数量
    protected final ProtocolHandler protocolHandler;
    ...
    public Connector(String protocol) {...}
}
```

9.6 创建 Tomcat 线程池

在 Tomcat 中根据业务配置不同，可以使用不同的参数创建线程池。

Tomcat 中的 Connector 将具体的协议处理托管给了不同的 ProtocolHandler 实现类。在协议处理器中有一个 AbstractEndpoint 成员对象。

```java
public abstract class AbstractProtocol<S>
            implements ProtocolHandler,MBeanRegistration {
    private final AbstractEndpoint<S,?> endpoint;
}
```

AbstractEndpoint 是一个抽象类，不同的协议需要提供不同的 Endpoint。这个类的作用就是提供底层的网络 I/O 的处理，不同的 ProtocolHandler 所内置的 Endpoint 不同，因此通过 AbstractEndpoint 为不同的协议处理类抽象出了一个实现框架。调用 createExecutor 方法，可以创建 Tomcat 线程池，参数信息都源于 9.5 节讲的<Executor>元素配置。

```java
public abstract class AbstractEndpoint<S,U> {
    public void createExecutor() {
        internalExecutor = true;
        TaskQueue taskqueue = new TaskQueue();
        TaskThreadFactory tf = new TaskThreadFactory(getName()
                        + "-exec-", daemon, getThreadPriority());
        executor = new ThreadPoolExecutor(getMinSpareThreads(),
                    getMaxThreads(), 60, TimeUnit.SECONDS,taskqueue, tf);
```

```
        taskqueue.setParent((ThreadPoolExecutor) executor);
    }
}
```

9.7　Web 服务器异步环境

AsyncContext 类是 Servlet 3.0 中新增的针对客户端请求的异步回应环境对象。Web 服务器在处理客户端请求时，对于一些不需要马上响应的请求，可以使用异步环境来处理，这样就可以大幅减少服务器的压力。

```
@WebServlet("/MyServlet")
public class MyServlet extends HttpServlet {
    protected void service(HttpServletRequest request, HttpServletResponse
                    response) throws ServletException, IOException {
        AsyncContext asy = request.startAsync();
    }
}
```

AsyncContext 对象创建后，接收 HTTP 请求的主线程立即结束。任务的处理和回应，交给异步环境来完成。Tomcat 有关 AsyncContext 的源代码参见如下：

```
public class Request implements HttpServletRequest {
    public AsyncContext startAsync(ServletRequest request,
                                    ServletResponse response) {
        //需要 Servlet 配置异步支持属性为 true
        if (!isAsyncSupported()) {
            IllegalStateException ise = new IllegalStateException(
                        sm.getString("request.asyncNotSupported"));
            throw ise;
        }
        if (asyncContext == null) {
            //创建 AsyncContext 对象
            asyncContext = new AsyncContextImpl(this);
        }
        asyncContext.setStarted(getContext(), request, response,
            request == getRequest() && response == getResponse().getResponse());
        asyncContext.setTimeout(getConnector().getAsyncTimeout());
        return asyncContext;
    } }
```

在异步环境实现类 AsyncContextImpl 中，使用多线程异步模式处理 HTTP 请求。

```
public class AsyncContextImpl implements AsyncContext,
```

```
                                        AsyncContextCallback {
    private volatile Request request;
    public AsyncContextImpl(Request request) {
        this.request = request;
    }
    public void start(final Runnable run) {
        if (log.isDebugEnabled()) {
            logDebug("start         ");
        }
        check();
        Runnable wrapper = new RunnableWrapper(run, context,
                            this.request.getCoyoteRequest());
        this.request.getCoyoteRequest().action(ActionCode.ASYNC_RUN,
                            wrapper);
    }
}
```

案例：AsyncContext 调用业务方法

测试步骤如下。

（1）在 JavaEE8 环境下，新建控制器（必须要设置异步支持模式）。

```
@WebServlet(urlPatterns="/LoginServlet",asyncSupported=true)
public class LoginServlet extends HttpServlet {
}
```

（2）调用 request 的 startAsync()方法，获得 AsyncContext。

```
protected void service(HttpServletRequest request,
                    HttpServletResponse response)
        throws ServletException, IOException {
    Log.logger.info("serivce begin : " + Thread.currentThread().getId());
    AsyncContext ac = request.startAsync();
    ...
}
```

（3）添加异步监听器。

```
ac.addListener(new AsyncListener() {
        @Override
        public void onTimeout(AsyncEvent arg0) throws IOException {
            Log.logger.info("超时结束:" + Thread.currentThread().getId());
        }
        @Override
```

```
        public void onStartAsync(AsyncEvent arg0) throws IOException {
            Log.logger.info("异步开始:" + Thread.currentThread().getId());
        }
        @Override
        public void onError(AsyncEvent arg0) throws IOException {
            Log.logger.info("异步错误:" + Thread.currentThread().getId());
        }
        @Override
        public void onComplete(AsyncEvent arg0) throws IOException {
            Log.logger.info("异步任务结束:"+Thread.currentThread().getId());
        }
});
```

（4）启动异步任务。

```
ac.start(new Runnable() {
    public void run() {
        UserBiz biz = new UserBiz();
        biz.m1();        //在异步环境中调用业务逻辑
        ac.complete();
    }
});
```

（5）模拟业务逻辑方法。

```
public class UserBiz {
    public void m1() {
        try {
            Thread.sleep(8000);
        } catch (Exception e) {
        }
        Log.logger.info("Userbiz:"+Thread.currentThread().getId());
    }
}
```

（6）在控制器的 service 方法中，输出主线程运行结果。

```
ac.getResponse().getWriter().println("hello:"
                            + Thread.currentThread().getId());
Log.logger.info("serivce end : " + Thread.currentThread().getId());
```

（7）通过浏览器两次访问 http://localhost:8080/UserWeb/LoginServlet，console 运行结果显示如下：

```
INFO - serivce begin : 16
```

```
INFO - serivce end : 16
INFO - Userbiz:17
INFO - 异步任务结束:18
INFO - serivce begin : 20
INFO - serivce end : 20
INFO - Userbiz:22
INFO - 异步任务结束:21
```

浏览器输出结果如下，注意：由于在业务逻辑方法 m1()中延时了 8 秒，浏览器上的显示结果也是在 8 秒后输出的。

```
hello:16
hello:20
```

9.8　Web 服务器 NIO

传统的 IO 模式是 BIO（Blocking I/O），即阻塞 IO 模式。从 JDK1.4 开始，引入了 NIO（NonBloking I/O）即非阻塞 IO 模式。Tomcat 既支持 BIO 模式，也支持 NIO 模式。

传统 BIO 是一种同步阻塞 IO 模式，即线程在进行 read()或 write()操作时，该线程无法进行其他操作，当 socket()通道中没有数据时，线程会一直处于阻塞等待状态，直到如下情况发生：①有数据可以读取；②数据读取完成；③发生异常。

NIO 有三大核心部分：Channel（通道）、Buffer（缓冲区）、Selector（多路复用器）。传统 BIO 基于字节流和字符流进行操作，而 NIO 基于 Channel 和 Buffer 进行操作。数据会从通道读取到缓冲区中，或者从缓冲区写入通道中（见图 9-2）。

Selector 用于监听多个通道的事件（如连接打开、数据到达等），因此单个线程可以监听多个数据通道（见图 9-3）。

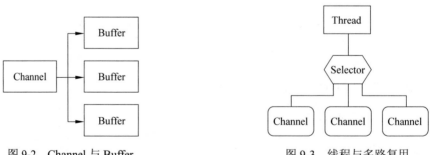

图 9-2　Channel 与 Buffer　　　　　　图 9-3　线程与多路复用

Java BIO 面向流意味着每次从流中读一个或多个字节，直至读取所有字节，它们没有被缓存在任何地方。此外，它不能前后移动流中的数据。如果需要前后移动从流中读取的数据，需要先将它缓存到一个缓冲区。NIO 的缓冲导向方法则不同，数据读取到一个它稍后处理的缓冲区，需要时可在缓冲区中前后移动。这就增加了处理过程中灵活性。但是，还需要检查是否该缓冲区中包含所有您需要处理的数据。而且，需确保当更多的数据读入缓冲区时，不要覆盖缓冲区里尚未处理的数据。

修改 Tomcat9 的 conf 目录下的 server.xml，可以对不同的连接器配置 NIO 协议模式。

```xml
<Connector port="8080" protocol="org.apache.coyote.http11.Http11Nio2Protocol"
           connectionTimeout="20000" redirectPort="8443" />
<Connector port="8009" protocol="org.apache.coyote.ajp.AjpNio2Protocol"
           redirectPort="8443" />
```

案例：服务器 NIO 处理请求

案例描述：模拟 Web 服务器采用 NIO 模式接收 HTTP 请求并回应的业务场景。
操作步骤如下。
（1）新建 Web 服务器类 WebServer，添加 start()方法如下：

```java
public void start() throws Exception {
    InetSocketAddress addr = new InetSocketAddress("127.0.0.1", 8080);
    ServerSocketChannel chanel = ServerSocketChannel.open();
    chanel.configureBlocking(false);              //配置非阻塞模式
    chanel.bind(addr);
    Selector selecor = Selector.open();           //打开一个新的选择器
    //选择器注册给通道
    chanel.register(selecor, SelectionKey.OP_ACCEPT);
    System.out.println("Web 服务器启动,端口号 8080...");
    while (true) {
        selecor.select();
        Iterator<SelectionKey> it = selecor.selectedKeys().iterator();
        while (it.hasNext()) {
            SelectionKey key = it.next();
            if (key.isAcceptable()) {
                SocketChannel socket = chanel.accept();
                socket.configureBlocking(false);
                socket.register(selecor,
                    SelectionKey.OP_READ | SelectionKey.OP_CONNECT);
            }
            if (key.isConnectable()) {
                SocketChannel socket = (SocketChannel) key.channel();
                socket.finishConnect();
            }
            if (key.isReadable()) {
                SocketChannel socket = (SocketChannel) key.channel();
```

```java
                ByteArrayOutputStream baos = new ByteArrayOutputStream();
                ByteBuffer buf = ByteBuffer.allocate(1024);//创建字节缓冲区
                while (socket.read(buf) > 0) {
                    buf.flip();
                    byte[] arr = buf.array();
                    baos.write(arr, 0, buf.limit());
                    buf.clear();
                }
                String msg = new String(baos.toByteArray());
                InetSocketAddress remoteAddr = (InetSocketAddress)
                socket.socket().getRemoteSocketAddress();
                String ip = remoteAddr.getAddress().getHostAddress();
                int port = remoteAddr.getPort();
                System.out.println("收到[" + ip + ":" + port + "]的请求:" + msg);
            }
            it.remove();
    }   }   }
```

（2）在主函数中启动 Web 服务器，系统提示如下：Web 服务器启动，端口号 8080...。

```java
public static void main(String[] args) {
    try {
        WebServer server = new WebServer();
        server.start();
    } catch (Exception e) {
        e.printStackTrace();
    }
}
```

（3）新建 WebClient 类，模拟客户端并发访问 Web 服务器。

```java
public static void main(String[] args) {
    ExecutorService pool = Executors.newFixedThreadPool(10);
    for(int i=0;i<10;i++) {
        pool.execute(new Runnable() {
            public void run() {
                try {
                    SocketChannel sc = SocketChannel.open();
                    InetSocketAddress addr
                            = new InetSocketAddress("127.0.0.1", 8080);
                    sc.connect(addr);
                    String msg = "hello , token=" + (Math.random()*100);
                    ByteBuffer buf
                            = ByteBuffer.wrap(msg.getBytes("ISO-8859-1"));
```

```
                    sc.write(buf);
                    System.out.println("客户端发送: " + msg);
                    sc.finishConnect();
                } catch (Exception e) {
                    e.printStackTrace();
                }
            }
        });
    } }
```

（4）执行 WebClient，模拟客户端发送并发的 HTTP 请求给 Web 服务器，程序运行结果如下：

```
客户端发送: hello , token=35.86021165382106
客户端发送: hello , token=62.842437681478266
客户端发送: hello , token=11.298099896820768
客户端发送: hello , token=53.06627692409256
客户端发送: hello , token=47.9250329455345
客户端发送: hello , token=64.91730553385268
客户端发送: hello , token=14.875847506738726
客户端发送: hello , token=28.32236525177577
客户端发送: hello , token=75.01098752006003
客户端发送: hello , token=14.124644776876094
```

（5）Web 服务器使用 NIO 模式接收客户端的 HTTP 请求。

```
收到[127.0.0.1:56841]的请求:hello , token=35.86021165382106
收到[127.0.0.1:56842]的请求:hello , token=62.842437681478266
收到[127.0.0.1:56845]的请求:hello , token=11.298099896820768
收到[127.0.0.1:56843]的请求:hello , token=53.06627692409256
收到[127.0.0.1:56847]的请求:hello , token=47.9250329455345
收到[127.0.0.1:56844]的请求:hello , token=64.91730553385268
收到[127.0.0.1:56846]的请求:hello , token=14.875847506738726
收到[127.0.0.1:56850]的请求:hello , token=28.32236525177577
收到[127.0.0.1:56849]的请求:hello , token=75.01098752006003
收到[127.0.0.1:56848]的请求:hello , token=14.124644776876094
```

9.9 本章习题

（1）关于 Tomcat 服务器的线程池描述错误的是（　　）。

A. Tomcat 服务器可以同时配置多个线程池

B. 部署在 Tomcat 服务器上的 Web 项目，尽量不要使用自定义的 ThreadPoolExecutor

C. Tomcat 服务器只能配置使用一个线程池

D. Tomcat 的线程池是 java.util.concurrent.ThreadPoolExecutor 的子类

（2）关于 Tomcat 服务器的任务队列描述正确的是（　　）。

A. Tomcat 中定义的阻塞队列 TaskQueue，它继承了 LinkedBlockingQueue

B. Tomcat 中定义的阻塞队列 TaskQueue，它继承了 ArrayBlockingQueue

C. Tomcat 中定义的阻塞队列 TaskQueue，它继承了 SynchronousQueue

D. Tomcat 的任务队列，可以使用 LinkedList

（3）关于 Tomcat 连接器 Connector 的描述，不正确的是（　　）。

A. Tomcat 可以同时配置多个连接器 Connector

B. 根据不同的客户端访问协议，可以配置不同的连接器 Connector

C. 不同的连接器 Connector，可以使用不同的线程池

D. 不同的连接器 Connector 分别对应不同的引擎和服务对象

（4）如下的哪个选项不属于 Connector 的参数？（　　）

A. protocol
B. maxThreads
C. maxConnections
D. executor
E. maximumPoolSize

第 10 章 并发编程应用

10.1 JVM 与多线程

Java 虚拟机是所有 Java 程序运行的基础，它与多线程密切相关。图 10-1 所示是 Hotspot 虚拟机的内存架构模型。

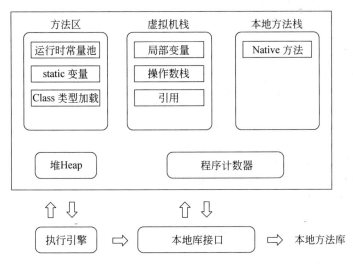

图 10-1 Hotspot 虚拟机的内存架构模型

"方法区"和"堆"属于多线程共享数据区。如 List aa = new ArrayList()，新创建的对象会默认创建在堆中，该对象允许多线程共享。

"虚拟机栈"属于线程私有数据区。如局部变量，在多线程环境中，每个局部变量都会进行独立的复制，线程之间数据互不影响。上述示例中的 aa 对象，如果是在方法内创建的，则 aa 也是局部变量。因此，要注意：aa 是引用，引用存放在虚拟机栈中，它是线程私有的。而堆中存储的是 aa 指向的 ArrayList()实例，它是多线程共享的。

"方法区"又叫静态区，存放所有的类型（class）、静态变量（static）、静态方法、运行时常量（如字符串常量）等内容。

Java 虚拟机栈（stack）的区域很小，特点是存取速度很快，所以在 stack 中存放的都是快速执行的任务，如基本数据类型的数据和对象的引用（reference）。

PC 寄存器（PC register）又称为程序计数器，每个线程启动的时候，都会创建一个 PC 寄存器。PC 寄存器里保存有当前正在执行的 JVM 指令的地址。每一个线程都有它自己的 PC 寄存器，它是该线程启动时创建的。PC 寄存器的内容总是指向下一条将被执行指令的地址，这里的地址可以是一个本地指针，也可以是在方法区中相对于该方法起始指令的偏移量。

"本地方法栈"中保存 Native 方法进入的地址。Native 方法是 Java 通过 JNI 直接调用本地 C/C++库。由于 Java 平台不允许直接使用指针，因此 JDK 中的底层操作都必须使用 Native 调用 C/C++代码实现（如多线程底层操作）。当线程调用 Java 方法时，虚拟机会创建一个栈帧并压入 Java 虚拟机栈。然而当它调用的是 Native 方法时，虚拟机会保持 Java 虚拟机栈不变，也不会向 Java 虚拟机栈中压入新的栈帧。

类的对象放在"堆（Heap）"中，所有的类对象都是通过 new 方法创建。对象创建后，在 stack（栈）中保存对象的引用，而对象的内存分配则是在堆中。

10.2 Servlet 与多线程

Servlet 默认为单例模式，因此在 Web 服务器中每个 Servlet 只有一个实例。在高并发的 Web 服务器环境中，Servlet 对象的 service()方法允许高并发访问。

Servlet 作为"单例+多线程"的工作模式，在开发中要非常小心。

Servlet 测试步骤如下：

（1）新建控制器 LoginServlet，用于用户登录控制。

```
@WebServlet("/LoginServlet")
public class LoginServlet extends HttpServlet {
    protected void service(HttpServletRequest request,
                    HttpServletResponse response)
                        throws ServletException, IOException {
        Log.logger.info("访问线程: " + Thread.currentThread().getId());
        Log.logger.info("Servlet 哈希:" + this.hashCode());
    }
}
```

（2）部署控制器 LoginServlet 到 Tomcat9 上，然后连续 5 次通过浏览器地址栏访问这个控制器：http://localhost:8080/UserWeb/LoginServlet。

程序运行结果如下，可以看出每次 HTTP 请求的处理线程 id 是不一样的，即 Servlet 是一个并发环境。而所有的 HTTP 请求，LoginServlet 的哈希值不变，说明 LoginServlet 为单例对象，不会因为 HTTP 请求次数的增加，出现多个 Servlet 对象。

```
INFO - 访问线程: 15
INFO - Servlet 哈希:6750462
```

```
INFO - 访问线程:16
INFO - Servlet 哈希:6750462
INFO - 访问线程:17
INFO - Servlet 哈希:6750462
INFO - 访问线程:18
INFO - Servlet 哈希:6750462
INFO - 访问线程:19
INFO - Servlet 哈希:6750462
```

（3）修改 LoginServlet 的代码如下，使用成员变量 uname 和 pwd 来接收客户端的请求参数（这是 struts2 框架的代码风格）。模拟高并发环境进行测试，必然会出现用户"张三"登录成功，而返回的用户名为"李四"的现象。这是因为，作为单例的 LoginServlet 对象，只能存储一个用户名和密码，后面请求的用户信息必然会替换掉前面的用户信息。因此，作为单例模式的 Servlet 对象，使用成员变量要非常小心。在 struts2 框架中，每个控制器都是 prototype 模式，因此使用成员变量接收用户信息是可行的。

```java
public class LoginServlet extends HttpServlet {
    private String uname;
    private String pwd;
    protected void service(HttpServletRequest request,
                           HttpServletResponse response)
                    throws ServletException, IOException {
        uname = request.getParameter("uname");
        pwd = request.getParameter("pwd");
        UserBiz biz = new UserBiz();
        User user = biz.login(uname, pwd);
        Log.logger.info("用户名:"+user.getUname() + "密码:" + user.getPwd());
    }
}
```

（4）再次修改 LoginServlet 的代码，用户信息作为局部变量，而 UserBiz 作为成员变量使用，想想这样的编码方式是否可行？

```java
public class LoginServlet extends HttpServlet {
    private UserBiz userBiz = new UserBiz();
    protected void service(HttpServletRequest request,
                           HttpServletResponse response)
                    throws ServletException, IOException {
        String uname = request.getParameter("uname");
        String pwd = request.getParameter("pwd");
        User user = userBiz.login(uname, pwd);
        Log.logger.info("用户名: " + user.getUname() + "密码: " + user.getPwd());
    }
}
```

分析：Servlet 是单例对象，它的成员变量 UserBiz 只有一次 new 的机会，因此不管 UserBiz 是否为单例，这里只能有一个 UserBiz 对象。

如果 UserBiz 没有使用成员变量，则大量的登录并发请求调用同一个 UserBiz 对象的 login()方法（这个方法中的变量都是局部变量），由于局部变量是线程隔离的，因此不会出错。

10.3 懒汉与恶汉模式

单例模式（Singleton Pattern）是最实用最简单的设计模式之一。它属于创建型模式，通过静态变量存储唯一的对象实例。在项目开发中，单例模式的应用非常普遍，如读取数据库连接的配置信息，只需要读取一次到单例对象中，以后的每次数据库连接都从单例中读取配置信息即可，这样性能会优化很多。

```java
public class Singleton {
    private static Singleton singleton;
    private Singleton() {
        //构造函数私有化，不允许在类外创建对象
        System.out.println(Thread.currentThread().getId() + "-创建对象");
    }
    public static Singleton instance() {
        if(singleton == null) {
            singleton = new Singleton();
        }
        return singleton;
    }
    public void m1() {
        try {
            Thread.sleep(100);
        } catch (Exception e) {
        }
    }
}
```

上面的代码是经典的单例模式写法，在普通环境中运行良好。但是在高并发环境中，可能会出现创建出多个对象实例的现象。参见如下测试代码：

```java
public static void main(String[] args) {
    ExecutorService pool = Executors.newCachedThreadPool();
    for(int i=0;i<3000;i++) {
        pool.execute(new Runnable() {
```

```java
        public void run() {
            Singleton st = Singleton.instance();
            st.m1();
        }
    });
}
pool.shutdown();
}
```

程序运行结果如下（每次效果不同），反复测试可以发现：上述 Singleton 代码，在高并发环境，会被创建出多个对象实例。

```
10-创建对象
12-创建对象
8-创建对象
17-创建对象
```

出现如上错误现象的原因是：多个线程同时调用 instance()方法时，判断 if(singleton == null) 时，都满足条件，因此会出现创建多个 Singleton 实例的现象。如何解决这个问题呢？

方案 1：最简单的解决方法就是加锁，如在 instance 方法前增加 synchronized 同步锁。这个方案最大的问题是：创建 Singleton 实例只有在第一次高并发调用时才可能出现错误，一旦 Singleton 实例已经生成，后面的 synchronized 就不会再起作用，而频繁的同步操作，势必会影响单例对象的后期调用性能。

```java
public synchronized static Singleton instance() {
    ...
}
```

方案 2：懒汉模式，通过同步锁与双重检查机制，既保证了不影响并发调用的性能，又确保了不会创建多余的单例对象。

```java
public static Singleton instance() {
    //此处检查不能用锁，防止影响性能
    if(singleton == null) {
        //线程同步，排队进入
        synchronized (Singleton.class) {
            //进入同步块后再次判断，保证所有排队线程只有一个可以创建实例
            if(singleton == null) {
                singleton = new Singleton();
            }
        }
    }
    return singleton;
```

方案 3：恶汉模式，在 Singleton 定义处即创建对象，无须等到 instance()方法再进行判断。这与在静态代码块中实例 Singleton 效果相同。这种模式，充分利用了 JVM 的底层机制，确保静态变量只会被实例一次。

```java
public class Singleton {
    private static Singleton singleton = new Singleton();
    ...
}
```

10.4　数据库 Connection 与多线程

在并发环境的操作中，为了使不同用户之间的数据访问互不影响，每个用户必须使用独立的关系型数据库 Connection，即每个用户线程拥有一个 Connection。

关系型数据库 Connection 是非常宝贵的物理资源，它受限于 CPU 等硬件条件。在并发性较高的应用系统中，数据库 Connection 的数量会远远小于在线用户的数量，因此 Connection 使用会出现排队等待现象，这通常会成为系统的性能瓶颈。

使用数据库 Connection 的基本原则为：晚打开、早关闭，让尽量少的 Connection 为更多的用户服务。

数据库 Connection 的使用模式基本有如下几种：Connection per method、Connection per logic、Connection per request 等。

1. Connection per method

在持久层的每个方法头打开数据库 Connection，在方法尾部的位置关闭数据库连接。这种使用模式最简单，但是问题也最多。首先就是数据库连接无法重用，如用户在淘宝提交订单，这会调用多个持久层的方法，如创建订单、生成订单明细、账户扣款、减少库存、生成积分等。如果每个持久层方法使用一个数据库 Connection，用完即马上关闭，在淘宝提交订单的操作中就会至少使用 5 个 Connection，这严重影响了性能。还有，如果 Connection 不能重用，是无法保证数据库的事务管理的（事务有 ACID 特性，交易中必须使用事务控制）。

```java
public class UserDao {
    public void regist(User user) throws Exception {
        //打开数据库 Connection
        ...
        //使用数据库 Connection
        ...
        //关闭数据库 Connection
    }
}
```

2. Connection per logic

参见图 10-2 所示的软件三层架构图,每个业务逻辑方法都可能会调用多个持久层方法,为了重用数据库 Connection、保证事务控制,合理的操作模式是在持久层的每个方法头打开数据库连接(每次打开时要判断当前线程中是否已有数据库连接,有则重用,没有就创建新连接),在服务层方法尾部统一关闭数据库连接。

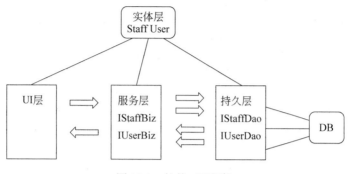

图 10-2 软件三层架构

3. Connection per request

在 Hibernate 框架应用中,由于 Hibernate 的懒加载特性,它要求在视图层调用懒加载对象信息时,可以访问数据库获取需要的信息。因此,数据库的连接应该在视图层关闭,而不是在常用的逻辑层关闭。即数据库的 Connection 在每一次的 HTTP 请求期间,都要保持有效。

如何实现数据库的 Connection 跨层传递?最有效的模式就是每个客户端请求线程分配一个数据库 Connection,当然这个线程必须要有数据库的访问才能分配数据库连接,如果单纯是访问静态资源或页面转向等操作,绝不能浪费 Connection。

使用 ThreadLocal 可以解决在客户线程中存储客户私有数据的需求,如存储数据库连接。

10.4.1 ThreadLocal 与线程私有数据

ThreadLocal 类可以存储与当前线程相关的私有数据,它的底层是 Map 结构。

```java
public class ThreadLocal<T> {
    public T get() {
        ...
    }
    public void set(T value) {
        ...
    }
    static class ThreadLocalMap {
        private Entry[] table;
        private int size = 0;
```

```java
        private static final int INITIAL_CAPACITY = 16;
        static class Entry extends WeakReference<ThreadLocal<?>> {
            Object value;

            Entry(ThreadLocal<?> k, Object v) {
                super(k);
                value = v;
            }
        }
    }
}
```

测试用 ThreadLocal 存取数据,操作步骤如下:

(1)创建静态变量 ThreadLocal。

```java
private static ThreadLocal<Integer> tl = new ThreadLocal<Integer>();
```

(2)主函数中,创建线程池,多线程并发操作向 ThreadLocal 中存入各自的数据。

```java
public static void main(String[] args) {
    ExecutorService pool = Executors.newFixedThreadPool(5);
    for(int i=0;i<5;i++) {
        pool.execute(new Runnable() {
            public void run() {
                int rand = (int)(Math.random() * 100);
                tl.set(rand);
                System.out.println(Thread.currentThread().getId()
                        + "存入: " + rand);
            }
        });
    }
    ...
}
```

(3)适当延迟后,用相同的线程池,每个线程分别读取自己存入的数据。

```java
public static void main(String[] args) {
    ExecutorService pool = Executors.newFixedThreadPool(5);
    ...
    try {
        Thread.sleep(200);
    } catch (Exception e) {
    }
    for(int i=0;i<5;i++) {
        pool.execute(new Runnable() {
```

```
        public void run() {
            Integer rand = tl.get();
            System.out.println(Thread.currentThread().getId()
                                    + "取出:" + rand);
            Thread.yield();
        }
    });
    }
    pool.shutdown();
}
```

程序运行结果如下:

8 存入：84
10 存入：50
12 存入：8
9 存入：19
11 存入：24
9 取出：19
11 取出：24
10 取出：50
11 取出：24
8 取出：84

10.4.2　ThreadLocal 存储数据库 Connection

在 ThreadLocal 中存储每个用户线程需要的数据库 Connection，然后该线程的所有数据库操作都使用这个 Connection，所有操作完毕，关闭这个数据库连接。

操作步骤如下：

（1）在工具类 DbFactory 中定义静态变量 ThreadLocal。

```
public class DbFactory {
    private static ThreadLocal<Connection> tl = new ThreadLocal<>();
}
```

（2）打开数据库连接（重用已有连接或创建新连接）。

```
public static Connection openConnection() throws Exception {
    Connection conn = tl.get();      //先查看是否有已打开的数据库连接
    if(conn == null || conn.isClosed()) {
        DbInfo db = DbInfo.instance();
        Class.forName(db.getDriver());
        conn = DriverManager.getConnection(db.getUrl(),
                             db.getUsername(), db.getPassword());
```

```java
            Log.logger.info(Thread.currentThread().getId() + "打开数据库...");
            tl.set(conn);                    //把已打开的数据库连接存入 tl
        }else {
            Log.logger.info(Thread.currentThread().getId()
                            + "使用已打开的数据库连接...");
        }
        return conn;
    }
```

（3）关闭数据库连接。

```java
    public static void closeConnection() {
        Connection conn = tl.get();
        if(conn != null) {
            try {
                conn.close();
                Log.logger.info(Thread.currentThread().getId()
                                + "关闭数据库连接...");
                tl.set(null);                //把 tl 中的 connection 置空
            } catch (Exception e) {
                Log.logger.error(e.getMessage(),e);
            }
        }
    }
```

（4）事务操作封装。

```java
    public static void beginTransaction() throws Exception{
        Connection conn = openConnection();
        if(conn != null) {
            conn.setAutoCommit(false);
            Log.logger.info(Thread.currentThread().getId() + "开启事务...");
        }
    }
    public static void commit() throws Exception{
        Connection conn = tl.get();
        if(conn != null) {
            conn.commit();
            Log.logger.info(Thread.currentThread().getId() + "提交事务...");
        }
    }
    public static void rollback() throws Exception{
        Connection conn = tl.get();
        if(conn != null) {
```

```java
        conn.rollback();
        Log.logger.info(Thread.currentThread().getId() + "回滚事务...");
    }
}
```

10.4.3　ThreadLocal 实现 Connection per logic 模式

模拟网上书城订单提交场景，在一次业务逻辑操作中分别要处理更新账户金额、生成订单、生成订单明细、修改图书库存等持久层操作。

在每个持久层方法的头部，都需要调用 DbFactory.openConnection()获取数据库连接，在逻辑层控制事务并关闭数据库连接。

（1）在持久层的 UserDao 中更新账户余额、创建新订单、添加订单明细。

```java
public class UserDao {
    //更新用户的账户余额
    public void updateUserAccount(String uname,double money) throws
                                        Exception{
        Connection conn = DbFactory.openConnection();
        ...
    }
    //创建新订单
    public String createNewOrder(String uname, double allMoney) throws
                                        Exception{
        Connection conn = DbFactory.openConnection();
        ...
    }
    //添加订单明细
    public void addOrderDetail(OrderDetail detail) throws Exception{
        Connection conn = DbFactory.openConnection();
        ...
    }
}
```

（2）在持久层的 BookDao 中更新图书数量。

```java
public class BookDao {
    //更新图书库存数量
    public void updateBookAccount(String isbn,int num) throws Exception{
        Connection conn = DbFactory.openConnection();
        ...
    }
}
```

(3)在逻辑层提交订单,控制事务,关闭数据库连接。

```java
public void buyBooks(String uname, double allMoney,
            Map<String, Integer> shopCar) throws Exception {
    UserDao userDao = new UserDao();
    BookDao bookDao = new BookDao();
    try {
        DbFactory.beginTransaction();       //开始事务
        userDao.updateUserAccount(uname, -allMoney);
        String orderNo = userDao.createNewOrder(uname, allMoney);
        for(String key : shopCar.keySet()) {
            //生成订单明细,写入库中
            ...
            userDao.addOrderDetail(detail);
        }
        DbFactory.commit();                 //无异常,就提交事务
    } catch (Exception e) {
        DbFactory.rollback();               //有异常,回滚
        Log.logger.error(e.getMessage(),e);
        throw e;
    } finally {
        DbFactory.closeConnection();        //关闭数据库连接
    }
}
```

10.4.4　ThreadLocal 实现 Connection per request 模式

对于 Connection per request 模式,可以使用 ServletRequestListener 监听器,在 requestDestroyed()方法中关闭数据库连接。每次 HTTP 请求,如果调用数据库,都会在持久层打开数据库连接,然后反复重用这个数据库连接,最后在 requestDestroyed()事件中关闭数据库连接。

```java
public class DbListener implements ServletRequestListener{
    public void requestInitialized(ServletRequestEvent sre) {
        ...
    }
    //每次 HTTP 请求结束时,调用 requestDestroyed 方法释放资源
    public void requestDestroyed(ServletRequestEvent sre) {
        DbFactory.closeConnection();        //关闭数据库连接
    }
}
```

10.5 高并发网站的 PageView 统计

在高并发网站中统计所有页面访问次数,这个问题通常被称为 PageView 统计。

实现高并发网站的 PageView 统计的操作步骤如下:

(1)定义 PageView 类,用静态变量存储所有页面的访问次数(也可以使用全局对象 ServletContext 实例)。为了防止并发统计时的冲突,此处使用 AtomicInteger 对象进行计数(不能使用 synchronized 同步锁,那样速度太慢)。

```java
public class PageView {
    private static AtomicInteger ac = new AtomicInteger(0);
    public static int addPageNum() {
        return ac.incrementAndGet();
    }
    public static int getPageViewNum() {
        return ac.get();
    }
}
```

(2)在过滤器中记录所有页面访问信息。

```java
@WebFilter(urlPatterns="/*")
public class PageFilter implements Filter{
    @Override
    public void doFilter(ServletRequest request,
                        ServletResponse response, FilterChain chain)
                    throws IOException, ServletException {
        HttpServletRequest req = (HttpServletRequest)request;
        Log.logger.info(Thread.currentThread().getId()
                    + "访问:" + req.getRequestURI());
        int pv = PageView.addPageNum();
        Log.logger.info("累计 PV 次数:" + pv);
    }
}
```

(3)通过浏览器访问网站地址,信息记录如下。

```
INFO - 489 访问:/BookShop/LoginServlet
INFO - 累计 PV 次数:1
INFO - 424 访问:/BookShop/MainSvl
INFO - 累计 PV 次数:2
INFO - 491 访问:/BookShop/MyHttpServlet
INFO - 累计 PV 次数:3
```

```
INFO - 341 访问：/BookShop/PicSvl
INFO - 累计 PV 次数：4
```

（4）使用 HttpURLConnection 对网站进行高并发访问，经过反复测试，上述代码运行正确，没有出现并发冲突现象。

```java
public static void main(String[] args) {
    ExecutorService service = Executors.newCachedThreadPool();
    for (int i = 0; i < 2000; i++) {
        service.execute(new Thread() {
            public void run() {
                try {
                    String url = "http://localhost:8081/BookShop";
                    HttpURLConnection conn = (HttpURLConnection) new
                            URL(url).openConnection();
                    conn.setRequestMethod("GET");
                    conn.setConnectTimeout(5000);
                    if(conn.getResponseCode() == 200){
                        System.out.println(Thread.currentThread().getId()
                                + "-请求成功...");
                    }
                } catch (Exception e) {
                    e.printStackTrace();
                }
            }
        });
    }
}
```

10.6 生成唯一的订单号

在购物网站，如当当、京东、淘宝、唯品会等，用户购买商品时都会创建唯一的订单号。在高并发系统，如淘宝的双十一零点抢购时，如何保证订单号唯一，同时还要确保系统性能不受影响呢？

订单号通常是有意义的数字，不能使用简单的时间戳来表示。下面示例生成如下格式的订单 D20190312-1234578，D 表示订单、20190312 为订单生成日期、1234578 为序列号。

操作步骤如下。

（1）新建工具类 OrderUtil，定义类型为 AtomicLong 的静态变量 al，用于在高并发环境下生成唯一的订单序列号。

```java
public class OrderUtil {
    private static AtomicLong al;
```

```
    ...
}
```

（2）在静态代码块中，初始化静态变量 al。

```
static{
    //万一服务器瘫痪，重启会导致订单号冲突，所以序号初始值从配置文件中读取
    //al = new AtomicLong(读取初始值+1);
    al = new AtomicLong(new Date().getTime());    //此为开发中的模拟数据
}
```

（3）生成唯一的订单号，格式为：D20190312-1234578。

```
public static String createNewOrderNo(){
    Calendar rightNow = Calendar.getInstance();
    int year = rightNow.get(Calendar.YEAR);
    int month = rightNow.get(Calendar.MONTH) + 1;
    String sMonth,sDay;
    if(month<10){
        sMonth = "0" + Integer.toString(month);
    }else{
        sMonth = Integer.toString(month);
    }
    int day = rightNow.get(Calendar.DAY_OF_MONTH);
    if(day<10){
        sDay = "0" + Integer.toString(day);
    }else{
        sDay = Integer.toString(day);
    }
    long xh = al.getAndIncrement();    //高并发环境下，流水号也不会冲突
    String dno = "D" + year + sMonth + sDay + "-" + xh;
    return dno;
}
```

（4）使用关系型数据库，如 MySQL 的 AUTO_INCREMENT 也可以生成唯一的订单号。MySQL 数据库可以保证在高并发环境下自增长的序列号不会冲突。

```
create table TOrder (
    xh              int(11) AUTO_INCREMENT,
    ono             varchar(30)         not null,
    allpay          double,
    constraint PK_ORDER primary key (xh)
);
insert into torder(ono,allpay)
        values(concat('d20190507-',last_insert_id()+1),1000);
```

（5）使用关系型数据库，如 Oracle 的 sequence 也可以生成唯一的订单号。Oracle 数据库可以保证在高并发环境下自增长的序列号不会冲突。

```
create table TOrder(
    ono             varchar(30)        not null,
    allpay          double,
    constraint PK_ORDER primary key (ono)
);
insert into torder values('D20190507-'||seq_id.nextval,1000);
```

10.7 浏览器并发请求限制

每一个客户端的 HTTP 请求，Web 服务器都需要使用一个独立的线程建立 socket 通道进行处理。当你访问百度首页时（https://www.baidu.com/），浏览器首先会接收 Web 服务器回应的第一个 HTML 页面。然后浏览器解析 HTML 中的静态资源或脚本，接着自动发送其他 HTTP 请求。如图 10-3 所示，一次百度首页的访问，浏览器会发送 73 个 HTTP 请求。为了减少 Web 服务器的并发压力，通常浏览器会限制并发访问同一个 Web 站点的线程数量。

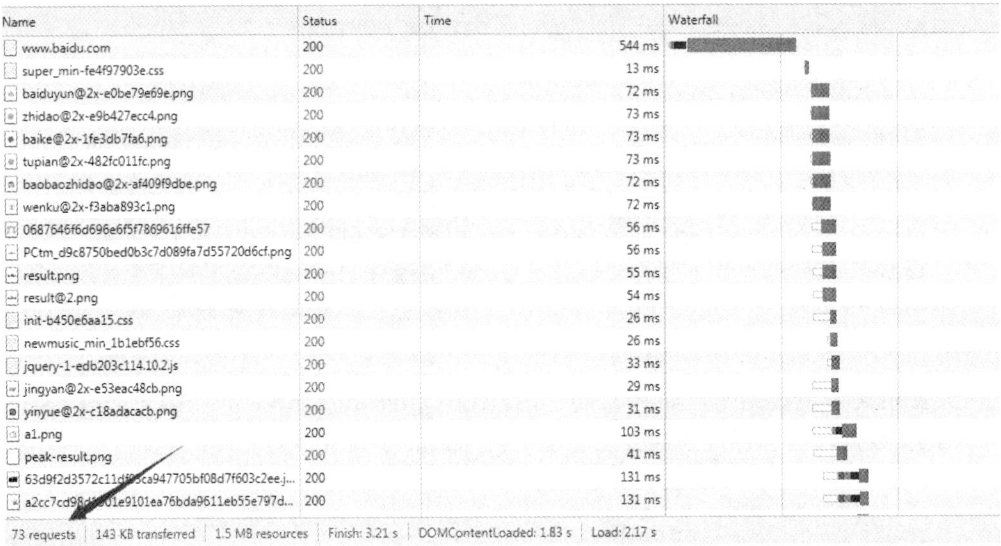

Name	Status	Time	Waterfall
www.baidu.com	200	544 ms	
super_min-fe4f97903e.css	200	13 ms	
baiduyun@2x-e0be79e69e.png	200	72 ms	
zhidao@2x-e9b427ecc4.png	200	73 ms	
baike@2x-1fe3db7fa6.png	200	73 ms	
tupian@2x-482fc011fc.png	200	73 ms	
baobaozhidao@2x-af409f9dbe.png	200	72 ms	
wenku@2x-f3aba893c1.png	200	72 ms	
0687646f6d696e6f5f7869616ffe57	200	56 ms	
PCtm_d9c8750bed0b3c7d089fa7d55720d6cf.png	200	56 ms	
result.png	200	55 ms	
result@2.png	200	54 ms	
init-b450e6aa15.css	200	26 ms	
newmusic_min_1b1ebf56.css	200	26 ms	
jquery-1-edb203c114.10.2.js	200	33 ms	
jingyan@2x-e53eac48cb.png	200	29 ms	
yinyue@2x-c18adacacb.png	200	31 ms	
a1.png	200	103 ms	
peak-result.png	200	41 ms	
63d9f2d3572c11d[illegible]ca947705bf08d7f603c2ee.j...	200	131 ms	
a2cc7cd98[illegible]01e9101ea76bda9611eb55e797d...	200	131 ms	

73 requests | 143 KB transferred | 1.5 MB resources | Finish: 3.21 s | DOMContentLoaded: 1.83 s | Load: 2.17 s

图 10-3　百度首页抓包

为了测试浏览器的并发访问限制，我们可以在 Web 服务器端故意设置延时回应，这样就可以清晰地看到并发限制，如果使用谷歌浏览器，浏览器访问同一 Web 站点时，每次只允许同时发送 6 个 HTTP 请求（见图 10-4）。

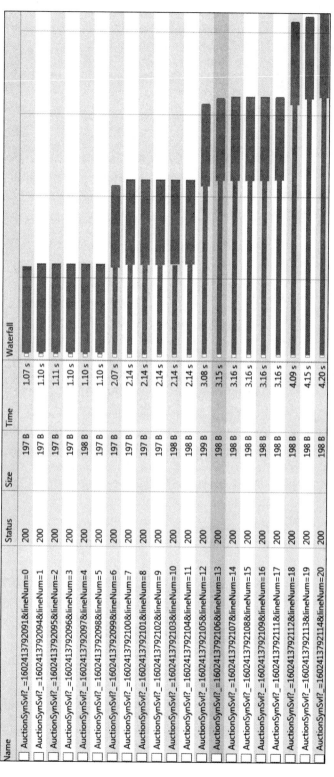

图10-4 浏览器并发请求限制

测试步骤如下：
（1）客户端使用 Jquery 模拟并发请求。

```
script language="javascript"
            src="<%=basePath%>script/jquery-1.4.4.min.js"></script>
```

（2）浏览器异步发送 HTTP 请求。

```javascript
function getMsg(i){
   var data="lineNum="+i;
   $.ajax({
       url: "AuctionSynSvl",
       type:"get",
       dataType:"html",
       data: data,
       timeout:50000,
       cache:false,
       success:function(data,testStatus) {
           var dataArray = Array();
           dataArray = data.split(" ");
           /*填充表格*/
           $("#table1").append("<tr id='tr"+i+"' class='mytr'></tr>");
           for(var j=0; j<dataArray.length; j++){
               $("#tr"+i).append("<td>"+dataArray[j]+"</td>");
           }
       }
   });
}
```

（3）客户端浏览器同时发送 30 个 HTTP 请求（这些请求会被浏览器分批发送出去）。

```javascript
function doGetMsg()    {
    $("#table1 tr.mytr").remove();
    for(var i=0; i<30; i++)
    {
        getMsg(i);
    }
}
```

（4）使用 Web 服务器的 Servlet 接收并处理浏览器的异步请求。

```java
public void service(HttpServletRequest request, HttpServletResponse response)
        throws ServletException, IOException {
    response.setContentType("text/html;charset=GBK");
    PrintWriter out = response.getWriter();
    try {
    Thread.sleep(1000);
```

```
            String name = Thread.currentThread().getName();
            long duration = System.currentTimeMillis();
            String linenum = request.getParameter("lineNum");
            out.println(linenum + " "+ name +" "+ duration);
            out.close();
        } catch (Exception e) {
            throw new RuntimeException(e.getMessage(), e);
        }
    }
```

（5）程序运行结果如图 10-5 所示，异步填充表格时，每次 6 个，分批完成。

userno	threadName	duration
		start
0	http-nio-8081-exec-4	1602413793145
1	http-nio-8081-exec-6	1602413793156
4	http-nio-8081-exec-10	1602413793156
3	http-nio-8081-exec-8	1602413793188
2	http-nio-8081-exec-7	1602413793188
5	http-nio-8081-exec-3	1602413793192
6	http-nio-8081-exec-9	1602413794194
7	http-nio-8081-exec-2	1602413794244
8	http-nio-8081-exec-1	1602413794248
9	http-nio-8081-exec-5	1602413794249
11	http-nio-8081-exec-6	1602413794250
10	http-nio-8081-exec-4	1602413794255
12	http-nio-8081-exec-10	1602413795201
13	http-nio-8081-exec-8	1602413795277
16	http-nio-8081-exec-9	1602413795281
14	http-nio-8081-exec-7	1602413795277
15	http-nio-8081-exec-3	1602413795280
17	http-nio-8081-exec-4	1602413795284
18	http-nio-8081-exec-1	1602413796215
19	http-nio-8081-exec-5	1602413796277
21	http-nio-8081-exec-4	1602413796294
23	http-nio-8081-exec-8	1602413796301
20	http-nio-8081-exec-6	1602413796303
22	http-nio-8081-exec-10	1602413796306
24	http-nio-8081-exec-7	1602413797228
25	http-nio-8081-exec-3	1602413797283
26	http-nio-8081-exec-9	1602413797314
27	http-nio-8081-exec-2	1602413797326
28	http-nio-8081-exec-1	1602413797340
29	http-nio-8081-exec-5	1602413797341

图 10-5　浏览器并发请求测试运行结果

10.8　NIO 与多路复用

参阅 9.8 节的 NIO 技术，多线程的多路复用技术在 Dubbo 架构、Redis 数据库、HTTP2 协议等很多业务场景都应用广泛。

图 10-6 是 Dubbo 的高层架构图。Registry 为服务的注册中心，Provider 是服务的提供者，Consumer 为服务的消费者。Consumer 与 Provider 之间是长链接，一个 Provider 允许很多 Consumer 同时并发访问。

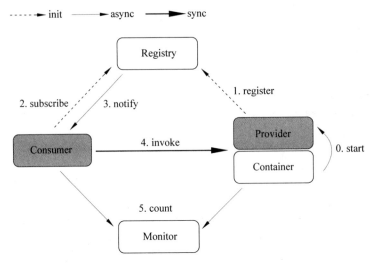

图 10-6　Dubbo 高层架构图

　　为了提高服务器的响应性能，解决 C10k 问题（即如果一个客户端请求，服务器分配一个响应线程，大量客户端请求会导致服务器线程上下文切换消耗过大，最终导致服务器系统崩溃），Dubbo 采用了著名的 NIO 框架 Netty 作为底层数据访问技术。

　　Netty 是一个高性能的、客户端/服务器模式的 NIO 框架。参见图 9-2 和图 9-3 的多线程多路复用模型，采用 Netty 框架，在服务器处理高并发请求时，通过多路复用技术可以使用很少的线程应付非常多的并发请求，从而缓解了服务器的压力。

　　Redis 是当前名气最大的内存数据库之一。Redis 性能非常高，如新浪微博的 Redis 集群，每天可以承受高达几千亿次的并发访问。Redis 采用的也是多线程的多路复用技术，服务器端只需要一个线程，就可以处理高达上百万的并发请求。当然这与 Redis 的数据存在内存中，数据结构非常简单，而且没有事务管理等因素有关。而 Dubbo 服务需要处理的是基于关系型数据库的事务操作，因此它的并发处理能力远逊于 Redis。

　　Dubbo 和 Redis 等多路复用框架，采用 NIO 模式处理并发请求时，应该注意所有请求任务为耗时短的轻量级任务，耗时长、数据量大的任务会严重影响 NIO 的并发处理性能，要坚决回避。

10.9　远程异步访问

　　在分布式系统中，客户端会调用远程服务，如 Dubbo Service、Web Service 等。调用远程服务的模式有同步调用和异步调用两种。同步调用就是远程服务收到请求后，在接收请求的线程中处理请求并返回结果；而异步调用的本质是远程服务收到请求后，在新的线程中处理请求，而接收请求的线程马上结束。

　　Dubbo Service 和 Web Service 都支持异步调用模式。本节通过 apache 的开源项目

HttpComponents 中的 HttpAsyncClient 来演示异步调用远程服务模式。

使用 HttpAsyncClient 调用远程资源的操作步骤如下所述。

（1）新建 Maven Project，并引入依赖配置。

```xml
<dependency>
    <groupId>org.asynchttpclient</groupId>
    <artifactId>async-http-client</artifactId>
    <version>2.8.1</version>
</dependency>
<dependency>
    <groupId>net.tascalate</groupId>
    <artifactId>net.tascalate.concurrent</artifactId>
    <version>0.8.0</version>
</dependency>
```

（2）使用 `AsyncHttpClient` 调用远程服务，并阻塞等待返回结果。

```java
private static void syncRequest() {
    String url = "http://www.oracle.com";
    AsyncHttpClient c = new DefaultAsyncHttpClient();
        Future<Response> f = c.prepareGet(url).execute();
        try {
          Response response = f.get();
            System.out.println(response.toString());
        } catch (Exception e) {
            e.printStackTrace();
        }
}
```

（3）在主函数中阻塞等待接收返回结果。

```java
public static void main(String[] args){
    System.out.println(Thread.currentThread().getId() + ",主线程启动...");
    syncRequest();
    System.out.println(Thread.currentThread().getId() + ",主线程结束...");
}
```

程序运行结果如下：

```
1,主线程启动...
NettyResponse {
    ...
}
1,主线程结束...
```

Future 的 get()方法是阻塞等待模式，因此调用线程一直在阻塞等待远程服务的返回结果。其实，AsyncHttpClient 的异步请求返回结果，还可以在异步响应事件中接收，操作步骤如下所述。

（1）创建 AsyncCompletionHandler，接收异步响应结果。

```java
private static void asyncRequest() {
    String url = "http://www.oracle.com";
    AsyncHttpClient c = new DefaultAsyncHttpClient();
    c.prepareGet(url).execute(
                    new AsyncCompletionHandler<Response>() {
        public Response onCompleted(Response response) throws Exception {
            System.out.println(response.toString());
            System.out.println(Thread.currentThread().getId()+"异步访问结束...");
            return response;
        }
    });
}
```

（2）在主函数中调用如下：

```java
public static void main(String[] args){
    System.out.println(Thread.currentThread().getId() + ",主线程启动...");
    asyncRequest();
    System.out.println(Thread.currentThread().getId() + ",主线程结束...");
}
```

异步响应事件的调用结果如下，主线程会先结束，然后才输出远程调用的结果。

```
1,主线程启动...
1,主线程结束...
NettyResponse {
    ...
}
10 异步访问结束...
```

上述代码使用异步响应事件模式，接收远程的访问结果。这样主线程无须阻塞等待远程调用的结果，响应事件的返回结果可以记录到某个位置。

对于远程异步访问，如果主线程希望阻塞等待返回结果，除了 Future 的 get()方法外，还可以使用 CountDownLatch 计数器。由于有些远程异步调用，并不会返回 Future，所以 CountDownLatch 的阻塞等待模式非常有用。参见如下代码，使用 CountDownLatch 计数器的 await()方法，阻塞等待远程返回的结果。

```java
private static void asyncRequest2() {
```

```java
        CountDownLatch cdl = new CountDownLatch(1);
        String url = "http://www.oracle.com";
            AsyncHttpClient c = new DefaultAsyncHttpClient();
            c.prepareGet(url).execute(
                            new AsyncCompletionHandler<Response>() {
              public Response onCompleted(Response response) throws Exception {
                 System.out.println(response.toString());
                   System.out.println(Thread.currentThread().getId()+"异步访问结束...");
                   cdl.countDown();
                   return response;
               }
            });
            try {
                cdl.await();
            } catch (InterruptedException e) {
                e.printStackTrace();
            }
    }
```

测试代码如下：

```java
public static void main(String[] args){
    System.out.println(Thread.currentThread().getId() + ",主线程启动...");
    asyncRequest2();
    System.out.println(Thread.currentThread().getId() + ",主线程结束...");
}
```

测试结果如下，当使用了 CountDownLatch 后，主线程调用 cdl.await()一直在阻塞等待异步响应事件 onCompleted()结束。当异步响应结束后，主线程的阻塞马上被打开。

```
1,主线程启动...
NettyResponse {
    ...
}
10 异步访问结束...
1,主线程结束...
```

10.10 防止缓存雪崩的 DCL 机制

假设如下业务场景：当当网的首页显示一批热点图书，这些图书对所有在线用户可见。如果每个用户访问当当网首页时，都从数据库加载热点图书数据，则数据库的压力会非常大。因此常用方案是把这些热点图书加载到 redis 中。

为了删除冷数据，还有及时刷新热点数据，redis 中的 cache 数据，通常会设置失效时间。

当 redis 中的 cache 数据失效后，会重新加载数据库中的数据到 redis 中。

假设当前存在 2000 个当当网首页并发请求，应该只允许一个线程读取数据库，其他请求都从 redis 中读取，这样系统的性能才有保障。如果 2000 个线程同时访问数据库读取热点图书数据，则会出现很多线程访问失败的现象。

为了实现上述需求，操作步骤如下所述。

（1）模拟从数据库读取所有热点图书数据。

```java
public class HotBookDao {
    private static Map<String,String> books;
    static {
        books = new HashMap<String,String>();
        books.put("is001", "isbn:9787302555301,bname:javaweb");
        books.put("is002", "isbn:9787302555302,bname:ssm");
        books.put("is003", "isbn:9787302555303,bname:三国演义");
        books.put("is004", "isbn:9787302555304,bname:水浒");
        books.put("is005", "isbn:9787302555305,bname:西游记");
    }
    public Map<String, String> getHotBooks() {
        return books;
    }
}
```

（2）读取热点图书，并加载到 redis 中。

```java
private void loadHotBook() {
    Jedis jedis = RedisUtil.getJedis();
    HotBookDao dao = new HotBookDao();
    Map<String,String> map = dao.getHotBooks();
    jedis.hmset("hotbook",map);
    jedis.expire("hotbook", 300); //每隔 5 分钟，重新加载一次
    RedisUtil.close(jedis);
    Log.logger.info(Thread.currentThread().getId()
            + "加载热点图书到 redis 中++++++");
}
```

（3）从 redis 中读取热点图书。

```java
private String getHotBookFromRedis(String isbn) {
    Jedis jedis = RedisUtil.getJedis();
    String value = jedis.hget("hotbook", isbn);
    Log.logger.info(Thread.currentThread().getId()
            + "从 redis 中读取热点图书...value=" + value);
    RedisUtil.close(jedis);
```

```
        return value;
    }
```

（4）定义重入锁。

```
public class HotBookService {
    private ReentrantLock lock = new ReentrantLock();
    ...
}
```

（5）当缓存失效时，通过 double check lock(DCL)机制加载数据到 redis 中。

```
public String readHotBook(String isbn) {
    String value = getHotBookFromRedis(isbn);
    if(value == null) {
        Log.logger.info(Thread.currentThread().getId() + "阻塞等待 锁...");
        lock.lock();
        Log.logger.info(Thread.currentThread().getId() + "获得锁----");
        //DCL 防止数据重复加载
        try {
            if((value=getHotBookFromRedis(isbn)) == null) {
                loadHotBook();                     //重新加载热点图书
                value = getHotBookFromRedis(isbn);//再次从 redis 中读取热点数据
            }
        } finally {
            lock.unlock();
        }
    }
    return value;
}
```

（6）通过日志观察数据的加载方式，模拟多用户的并发访问。

```
public static void main(String[] args) {
    HotBookService hb = new HotBookService();
    CountDownLatch cdl = new CountDownLatch(10);
    ExecutorService pool = Executors.newCachedThreadPool();
    for(int i=0;i<10;i++) {
        pool.execute(new Runnable() {
            public void run() {
                cdl.countDown();
                try {
                    cdl.await();
                } catch (InterruptedException e) {
```

```
            }
            hb.readHotBook("is002");
        }
    });
}
pool.shutdown();
```

程序运行结果如下，当缓存失效时，只有一个线程会访问数据库加载热点图书数据，其他线程都会从 redis 中读取数据。

```
INFO com.icss.util.Log - 12 从 redis 中读取热点图书...value=null
INFO com.icss.util.Log - 14 从 redis 中读取热点图书...value=null
INFO com.icss.util.Log - 15 从 redis 中读取热点图书...value=null
INFO com.icss.util.Log - 13 从 redis 中读取热点图书...value=null
INFO com.icss.util.Log - 16 从 redis 中读取热点图书...value=null
INFO com.icss.util.Log - 17 从 redis 中读取热点图书...value=null
INFO com.icss.util.Log - 12 阻塞等待锁...
INFO com.icss.util.Log - 8 从 redis 中读取热点图书...value=null
INFO com.icss.util.Log - 14 阻塞等待锁...
INFO com.icss.util.Log - 15 阻塞等待锁...
INFO com.icss.util.Log - 13 阻塞等待锁...
INFO com.icss.util.Log - 16 阻塞等待锁...
INFO com.icss.util.Log - 11 从 redis 中读取热点图书...value=null
INFO com.icss.util.Log - 12 获得锁----
INFO com.icss.util.Log - 10 从 redis 中读取热点图书...value=null
INFO com.icss.util.Log - 17 阻塞等待锁...
INFO com.icss.util.Log - 8 阻塞等待锁...
INFO com.icss.util.Log - 9 从 redis 中读取热点图书...value=null
INFO com.icss.util.Log - 11 阻塞等待锁...
INFO com.icss.util.Log - 10 阻塞等待锁...
INFO com.icss.util.Log - 9 阻塞等待锁...
INFO com.icss.util.Log - 12 从 redis 中读取热点图书...value=null
INFO com.icss.util.Log - 12 加载热点图书到 redis 中++++++
INFO com.icss.util.Log - 12 从 redis 中读取热点图书...value=isbn:9787302555302,bname:ssm
INFO com.icss.util.Log - 14 获得锁----
INFO com.icss.util.Log - 14 从 redis 中读取热点图书...value=isbn:9787302555302,bname:ssm
INFO com.icss.util.Log - 15 获得锁----
INFO com.icss.util.Log - 15 从 redis 中读取热点图书...value=isbn:9787302555302,bname:ssm
INFO com.icss.util.Log - 13 获得锁----
```

```
INFO com.icss.util.Log - 13 从 redis 中读取热点图书...value=isbn:9787302555302,
bname:ssm
INFO com.icss.util.Log - 16 获得锁----
INFO com.icss.util.Log - 16 从 redis 中读取热点图书...value=isbn:9787302555302,
bname:ssm
INFO com.icss.util.Log - 17 获得锁----
INFO com.icss.util.Log - 17 从 redis 中读取热点图书...value=isbn:9787302555302,
bname:ssm
INFO com.icss.util.Log - 8 获得锁----
INFO com.icss.util.Log - 8 从 redis 中读取热点图书...value=isbn:9787302555302,
bname:ssm
INFO com.icss.util.Log - 10 获得锁----
INFO com.icss.util.Log - 10 从 redis 中读取热点图书...value=isbn:9787302555302,
bname:ssm
INFO com.icss.util.Log - 9 获得锁----
INFO com.icss.util.Log - 9 从 redis 中读取热点图书...value=isbn:9787302555302,
bname:ssm
INFO com.icss.util.Log - 11 获得锁----
INFO com.icss.util.Log - 11 从 redis 中读取热点图书...value=isbn:9787302555302,
bname:ssm
```

10.11 分布式锁解决商品超卖

在分布式系统中（如京东、淘宝），为了满足高并发需求，业务逻辑会部署在应用服务器集群中（如 dubbo 集群）。前面我们讲过的很多锁，如 synchronized 隐式锁和 reentrantlock 重入锁都是基于 JVM 的，即这些锁只能在单台服务器上有效，而在集群环境就可能会出现问题。这时，使用脱离 JVM 的分布式锁解决线程排队问题，就非常有必要了。

分布式锁是一个概念，用于解决分布式环境下的线程排队问题。常用的方案有基于 redis 的 Redlock 方案和基于关系型数据库的锁方案等。

下面模拟淘宝双 11 商品抢购，来演示分布式锁的应用效果。操作步骤如下：

（1）淘宝平台准备了多种商品，在双 11 零点进行抢购。下面代码模拟了多种商品，并设置了抢购数量。

```java
public class GoodsDao {
    private static Map<String,String> map;
    static {
        map = new HashMap<String,String>();
        map.put("g001", "200");
        map.put("g002", "100");
        map.put("g003", "150");
        map.put("g004", "300");
```

```java
        map.put("g005", "100");
    }
    public Map<String, String> getMap() {
        return map;
    }
}
```

（2）为了解决商品的超卖问题，商品抢购数据都加载到 redis 中：

```java
public class GoodService {
    public void init() {
        GoodsDao dao = new GoodsDao();
        Map<String,String> map = dao.getMap();
        Jedis jedis = RedisUtil.getJedis();
        jedis.hmset("goodList",map);
        RedisUtil.close(jedis);
        Log.logger.info(Thread.currentThread().getId() + "加载拍卖商品...");
    }
}
```

（3）利用 redis 的 SET 指令，进行线程排队。SET 指令的格式如下。

```
SET  key  value  [EX seconds]  [PX milliseconds]  [NX|XX]
```

当 SET 指令设置了 NX 参数，或直接使用 SETNX 指令，可以使 SET 指令具有原子性，即只有指定的 key 不存在时，才会设置成功并返回 1，否则设置失败返回 0。即使在高并发环境，redis 也能保证，只有一条 SEXNX 指令设置成功。

因为 redis 是独立于应用服务集群的，所以使用 redis 的 SETNX 指令，可以起到分布式锁的效果。

```java
public class GoodService {
    private boolean isLeft = true;              //判断商品是否有剩余
    private void createOrder(String gno) {
        Log.logger.info(Thread.currentThread().getId() + "创建订单,商品编号=" + gno);
    }
    ...
}
```

（4）在 GoodService 中定义商品抢购的方法 auction()。

```java
public void auction(String gno) {
    Log.logger.info(Thread.currentThread().getId() + "抢拍: " + gno);
    Jedis jedis = RedisUtil.getJedis();
    //多线程抢锁,只能有一个线程 setnx()返回1,其他线程循环等待
    while(jedis.setnx(gno, "lock")==0) {
```

```java
        }
        Log.logger.info(Thread.currentThread().getId() + "获取锁...");
        if(isLeft) {
            jedis.expire(gno, 5);        //防止意外出现死锁,必须要设置 key 的失效时间
            String value = jedis.hget("goodList", gno);
            Integer iValue = Integer.parseInt(value);
            if(iValue>0) {
                iValue = iValue -1;
                jedis.hset("goodList", gno, iValue.toString());
                Log.logger.info(Thread.currentThread().getId()
                        + "抢到商品:" + gno + ",剩余:" + iValue);
                createOrder(gno);        //生成订单
            }else {
                Log.logger.info(Thread.currentThread().getId()
                        + "," + gno + "已售完...");
                isLeft = false;
            }
        }else {
            Log.logger.info(Thread.currentThread().getId()
                    + "商品已售完,抢购失败------");
        }
        jedis.del(gno);        //确保要删除 key,这样其他线程才能获得锁
        jedis.close();
    }
```

定义成员变量 isLeft,判断商品是否已售完非常重要。确保所有排队抢购的线程,在商品已售完时快速结束,不要影响其他线程的处理。

(5)模拟多用户抢购商品的场景。

```java
public static void main(String[] args) {
    GoodService service = new GoodService();
    service.init();
    CountDownLatch cdl = new CountDownLatch(2000);
    ExecutorService pool = Executors.newCachedThreadPool();
    for(int i=0;i<2000;i++) {
        pool.execute(new Runnable() {
            public void run() {
                cdl.countDown();
                try {
                    cdl.await();
                } catch (InterruptedException e) {
                }
                service.auction("g002");
```

```
                }
            });
        }
        pool.shutdown();
    }
```

如果商品 g002 的可抢购数量为 10，排队抢购的用户为 20 人，则程序的运行效果如下：

```
11:00:54.463 INFO com.icss.util.Log - 1 加载拍卖商品...
11:00:54.463 INFO com.icss.util.Log - 11 抢拍：g002
11:00:54.463 INFO com.icss.util.Log - 14 抢拍：g002
11:00:54.463 INFO com.icss.util.Log - 11 获取锁...
11:00:54.463 INFO com.icss.util.Log - 13 抢拍：g002
11:00:54.478 INFO com.icss.util.Log - 11 抢到商品：g002，剩余：9
11:00:54.478 INFO com.icss.util.Log - 11 创建订单，商品编号=g002
11:00:54.478 INFO com.icss.util.Log - 12 抢拍：g002
11:00:54.478 INFO com.icss.util.Log - 17 抢拍：g002
11:00:54.510 INFO com.icss.util.Log - 15 抢拍：g002
11:00:54.510 INFO com.icss.util.Log - 16 抢拍：g002
11:00:54.510 INFO com.icss.util.Log - 30 抢拍：g002
11:00:54.510 INFO com.icss.util.Log - 18 抢拍：g002
11:00:54.510 INFO com.icss.util.Log - 19 抢拍：g002
11:00:54.510 INFO com.icss.util.Log - 21 抢拍：g002
11:00:54.510 INFO com.icss.util.Log - 20 抢拍：g002
11:00:54.510 INFO com.icss.util.Log - 22 抢拍：g002
11:00:54.510 INFO com.icss.util.Log - 23 抢拍：g002
11:00:54.510 INFO com.icss.util.Log - 24 抢拍：g002
11:00:54.510 INFO com.icss.util.Log - 25 抢拍：g002
11:00:54.525 INFO com.icss.util.Log - 26 抢拍：g002
11:00:54.525 INFO com.icss.util.Log - 27 抢拍：g002
11:00:54.525 INFO com.icss.util.Log - 28 抢拍：g002
11:00:54.525 INFO com.icss.util.Log - 29 抢拍：g002
11:00:55.009 INFO com.icss.util.Log - 23 获取锁...
11:00:55.009 INFO com.icss.util.Log - 23 抢到商品：g002，剩余：8
11:00:55.009 INFO com.icss.util.Log - 23 创建订单，商品编号=g002
11:00:55.509 INFO com.icss.util.Log - 13 获取锁...
11:00:55.509 INFO com.icss.util.Log - 13 抢到商品：g002，剩余：7
11:00:55.509 INFO com.icss.util.Log - 13 创建订单，商品编号=g002
11:00:56.024 INFO com.icss.util.Log - 25 获取锁...
11:00:56.024 INFO com.icss.util.Log - 25 抢到商品：g002，剩余：6
11:00:56.024 INFO com.icss.util.Log - 25 创建订单，商品编号=g002
11:00:56.524 INFO com.icss.util.Log - 15 获取锁...
11:00:56.524 INFO com.icss.util.Log - 15 抢到商品：g002，剩余：5
11:00:56.524 INFO com.icss.util.Log - 15 创建订单，商品编号=g002
```

```
11:00:57.039 INFO com.icss.util.Log - 14 获取锁...
11:00:57.039 INFO com.icss.util.Log - 14 抢到商品：g002，剩余：4
11:00:57.039 INFO com.icss.util.Log - 14 创建订单,商品编号=g002
11:00:57.570 INFO com.icss.util.Log - 17 获取锁...
11:00:57.570 INFO com.icss.util.Log - 17 抢到商品：g002，剩余：3
11:00:57.570 INFO com.icss.util.Log - 17 创建订单,商品编号=g002
11:00:58.085 INFO com.icss.util.Log - 21 获取锁...
11:00:58.085 INFO com.icss.util.Log - 21 抢到商品：g002，剩余：2
11:00:58.085 INFO com.icss.util.Log - 21 创建订单,商品编号=g002
11:00:58.585 INFO com.icss.util.Log - 30 获取锁...
11:00:58.58 INFO com.icss.util.Log - 30 抢到商品：g002，剩余：1
11:00:58.585 INFO com.icss.util.Log - 30 创建订单,商品编号=g002
11:00:59.100 INFO com.icss.util.Log - 28 获取锁...
11:00:59.100 INFO com.icss.util.Log - 28 抢到商品：g002，剩余：0
11:00:59.100 INFO com.icss.util.Log - 28 创建订单,商品编号=g002
11:00:59.600 INFO com.icss.util.Log - 24 获取锁...
11:00:59.600 INFO com.icss.util.Log - 24,g002 已售完...
11:00:59.600 INFO com.icss.util.Log - 26 获取锁...
11:00:59.600 INFO com.icss.util.Log - 26 商品已售完，抢购失败------
11:00:59.600 INFO com.icss.util.Log - 19 获取锁...
11:00:59.600 INFO com.icss.util.Log - 19 商品已售完，抢购失败------
11:00:59.600 INFO com.icss.util.Log - 22 获取锁...
11:00:59.600 INFO com.icss.util.Log - 22 商品已售完，抢购失败------
11:00:59.600 INFO com.icss.util.Log - 12 获取锁...
11:00:59.600 INFO com.icss.util.Log - 12 商品已售完，抢购失败------
11:00:59.600 INFO com.icss.util.Log - 20 获取锁...
11:00:59.600 INFO com.icss.util.Log - 20 商品已售完，抢购失败------
11:00:59.600 INFO com.icss.util.Log - 27 获取锁...
11:00:59.600 INFO com.icss.util.Log - 27 商品已售完，抢购失败------
11:00:59.600 INFO com.icss.util.Log - 29 获取锁...
11:00:59.600 INFO com.icss.util.Log - 29 商品已售完，抢购失败------
11:00:59.600 INFO com.icss.util.Log - 18 获取锁...
11:00:59.600 INFO com.icss.util.Log - 18 商品已售完，抢购失败------
11:00:59.600 INFO com.icss.util.Log - 16 获取锁...
11:00:59.600 INFO com.icss.util.Log - 16 商品已售完，抢购失败------
```

  使用 redis 的分布式锁与 Reentrantlock 重入锁相比，还有一个优势，就是 SETNX 只锁定某个商品的 key，这不会影响到其他商品的抢购。redis 分布式锁的粒度比 Reentrantlock 重入锁要更细，性能更高。

  如果商品抢购数据放在 MySQL 中，用乐观锁、共享锁、排他锁保证商品不会超卖，这也是分布式锁的解决方案之一。但是关系型数据库的 connection 数量非常有限，在高并发商品抢购的场景下，关系型数据库的压力会非常大，性能远远不如 redis 的 Redlock 方案。

参 考 文 献

[1] Bruce Eckel .Thinking in Java[M]. 陈昊鹏，译. 北京：人民邮电出版社，2006.
[2] Brian Goetz. Java Concurrency in Practice[M]. 韩锴，方妙，译. 北京：电子工业出版社，2007.
[3] Doug Lea.The java.util.concurrent Synchronizer Framework[EB/OL].[2020-10-20]. http://gee.cs.oswego.edu/dl/papers/aqs.pdf.